# WHAT
# WE VALUE

# WHAT
# WE VALUE

## THE NEUROSCIENCE OF
## Choice and Change

*Emily Falk*

**W. W. NORTON & COMPANY**

*Independent Publishers Since 1923*

For information about permission to reproduce selections from this
book, write to Permissions, W. W. Norton & Company, Inc.,
500 Fifth Avenue, New York, NY 10110

For information about special discounts for bulk purchases, please
contact W. W. Norton Special Sales at specialsales@wwnorton.com or
800-233-4830

All diagrams are by Omaya Torres.

Manufacturing by Lakeside Book Company
Book design by Brooke Koven
Production manager: Louise Mattarelliano

ISBN 978-1-324-03709-5

W. W. Norton & Company, Inc.
500 Fifth Avenue, New York, NY 10110
www.wwnorton.com

W. W. Norton & Company Ltd.
15 Carlisle Street, London W1D 3BS

10 9 8 7 6 5 4 3 2 1

*For my mom, Katherine, and my grandma, Beverly*

# CONTENTS

# INTRODUCTION

IT WAS A TYPICAL night at my house. One of my kids was jumping on the arm of the couch, playing a new song he had learned on the guitar. The other was trying to show my grandma, Bev, a new Lego creation, proudly thrusting the bricks within inches of her eyeballs. As dishes clanged and my phone pinged, I felt a now-familiar sense of constriction—like being stuck in a cave where the walls are closing in.

Many of us know this feeling—of being trapped in an impossible negotiation of trade-offs, of needing to choose between different things that matter in different ways. Whether it's supporting colleagues at work or trying to protect a weekend of quality time with family, it never seems like there's enough of me to choose everything I want to choose.

On that night, I looked at Bev—one of my favorite people in the world—and chose to prioritize some quality time with her. I took her hand and guided her around piles of cast-off Legos, past the remains of a wooden block castle in disarray, and out the door.

At ninety-nine, Bev's hands are soft and strong, and I try to memorize them, how thin her skin feels, like smooth tissue paper, gripping mine. Outside, I could breathe again. I felt the momentary relief that comes with having made what you think is the right choice. But only briefly—only until Bev turned to me and said that although she liked coming to my house and seeing my kids, we weren't *really* spending time together.

I dropped her hand. "Of course we are," I insisted.

"Not really," she said. "We get to see each other, but when I come to your house, you're not really paying attention to me." I knew she was really saying, though not aloud, "I see that you think we're spending quality time together, but we can do better, go deeper, than a ten-minute walk around the block."

I didn't want Bev to be right, but I knew that she was. In the back of my mind, there's often a whisper reminding me to spend more quality time with her, but it's a whisper among so many shouts. When she proposed changing our routine so that I came to her house instead of she to mine, the shouts started up again: kids, work, the traffic and parking situation near her apartment. Under the open night sky, I found myself back in that cave, elbows pushed into my sides, shoulders pressed up to my ears. How could I navigate through this feeling?

Maybe you've also been in a situation like this—one where you know that there's an important thing you need to do, but you can't seem to do it. Maybe your doctor is concerned about your health, and you know you need to exercise more, but in what little free time you have it's hard not to sink into a favorite TV show. Maybe you've been wanting to make more time to mentor a promising person on your team at work, but urgent deadlines keep you from getting there. Maybe you have a goal to meet new people, but you find yourself talking with the same familiar friends or, worse, staring at your phone every time you go to a party or event.

The structure of the situation is familiar to so many of us: I want to do the thing, and the thing is important to me, but it's also *hard*. For . . . *reasons.*

Though I didn't think about it this way at the time, this is the same basic problem I've studied for most of my career: how we choose—including how we choose to change. Every morning, I walk to the University of Pennsylvania, where I direct Penn's Communication

Neuroscience Lab, and my team and I design experiments to explore (among other things) the relationship between what people value, the choices they make, and how this is shaped by the outside world. Specifically, we use neuroimaging to explore the brain systems that handle this process, and in doing so we helped discover how these systems relate to the ways people spend their time, change their behavior, and connect with others. So shouldn't I be the expert? Bev is one of the most important people in my life. Shouldn't I know how to make a choice to prioritize time with her? Shouldn't I be in control of what's valuable to me?

It seems I wasn't. It was hard even to clear out enough space in my defenses to pause before telling her she was wrong, let alone to ask myself: *What's going on here? Why am I resisting visiting one of my favorite people?*

*Why is* this *the choice I'm making?*

And worse, *Why do I* keep *making this choice, over and over?*

If a friend had posed this dilemma to me, I might have told them that we often focus so much on the results of a choice that we miss the opportunity to understand *why* we made it to begin with, making lasting change harder. One way to adjust that kind of thinking is to understand a system in the brain that's fundamental to many of the choices we make. Neuroscientists like me call it the *value system.*

People are sometimes surprised to hear a neuroscientist talk about a "value system" and "what we value." When they think of "values," they might think of moral values—a code of conduct, a sense of what is intrinsically good and right, or a few important principles we choose to live by. Alternatively, they might think of economists or market analysts discussing prices or the feeling of getting a good deal at the store. But when neuroscientists talk about value, we mean, most basically, the amount of reward your brain expects you to derive from a particular action in a particular moment.

With every choice we make, the value system's job is to weigh disparate elements against each other in what my colleagues and I call the *value calculation*. These elements indeed include things like moral values and the economic value of an option, but they also include the consequences of your past choices, your mood, the opinions of the people around you, and so much more. A reward can be money, but it can also be friendship. It can be seeing something good happen in the world for others, achieving a small goal, or having enough energy and strength to finally run a marathon. There are many things that our brains value, many ways our brains can find reward—but as we find ourselves making the same choices again and again, it doesn't always feel that way. Getting takeout trumps saving for retirement; hitting deadlines trumps professional development; the Internet vortex trumps spending time with the people we love. In this way, the choices the brain hands down don't always align with what we might explicitly think of as the thing we value most.

Sometimes this is because external expectations are unreasonable, but sometimes it's within our control to make a different choice. And the value system is at the heart of these decisions to change, too. I began my career in the late aughts and early 2010s looking at what happens in people's brains when they choose to change their behavior. In a series of experiments, my graduate school adviser, Matt Lieberman, fellow graduate student Elliot Berkman, and I scanned people's brains as they responded to messages about wearing sunscreen and quitting smoking. After I became a professor, we continued with similar experiments encouraging people to exercise more and drive safely. Our goal was to identify what was going on inside people's brains as they considered how they might change, and then to see if they actually did. Back then, no one knew if it would be possible to link what we saw in a neuroimaging lab to actual behavior change. But when we started to see a pattern in the data, we realized that we had identified an important intervention point, one we could target to help people change.

We found that if parts of a person's value system, like a region known as the medial prefrontal cortex, ramped up their activity when they saw a message about sunscreen or smoking or exercise, they were more likely to change their behavior to conform to the message—regardless of whether they said they consciously thought the message was effective. This offered our first glimpse of how the value system was linked to relatively high-stakes, real-life choices outside the lab. A plethora of other studies, by my team and others, have shown similar findings when people are deciding what to eat, what to buy, how much to save for retirement, and more.

At first we were only looking to see if activity in the brain correlated with the choices people made outside the lab. Once we saw that it did, we asked: *How can we use this to help facilitate change?* I believed that the answer was to somehow ramp up activity in the system, but it would take more than a decade of research to understand how.

During that time, in experiments ranging from giving people feedback about their peers' experiences, to helping them connect with their core values as a way of becoming more open to change, to comparing how the value system responds to immediate rewards versus those that lie in the more distant future, my team and others saw how simple interventions could dial value system activity up or down, which could ultimately help someone change their behavior. We discovered how changing where people put their attention—on different past experiences, current needs, or dreams for the future—*changes* the value calculation. This research also made it clear that activity in the value system captures something that goes beyond people's initial instincts about what they'll do next and can sometimes help explain the discrepancy we observed between what people say they will do and what they actually do.

As research on the value system progressed, we learned that the value system isn't only measuring what we think we *should* do in the abstract, or what we'd *want* to do if we were our best selves. There is so much more going on under the surface than the basic push and pull between desire and reason. The value system takes into account what we've done before and what the outcomes were. It asks: What do

I need, *right now?* The solution isn't simply to try harder, to will ourselves to make "good" decisions so that our self-control can override our baser impulses. When we understand how and why our brains make decisions, it highlights different inputs to the value calculation that we might focus on to shape the choices we make and how we feel about them. This reveals new potential intervention points, and each of those can represent an opportunity for change.

In this way, I like to think of understanding the value system as a means of having a flashlight in the cave—one that helps us gain clarity about what shapes choices for ourselves and others. My team and others have found that being clear about what we want and why is a key ingredient for happiness and well-being, but that people vary a lot in how much they tend to know why they are doing what they are doing. This understanding might make us more compassionate toward ourselves and each other, showing us that there are *reasons* we make the choices we do, even if our best selves might make a different choice or we wish we'd done something different in retrospect. But even beyond this compassion (which I'd argue can itself be transformative), this understanding can help us make *different* choices, maybe aligning our daily decisions better with our big-picture goals and values. Shining a flashlight around a dark cave might reveal a pulley that opens a door or a lever that reveals a skylight. Sometimes there are whole new pathways that we didn't know were there—they just weren't illuminated. If we know how the inner workings operate, it becomes easier to understand ourselves and others and to better navigate our way through, together.

As for me, I kept thinking back to what Bev had said. I had known for a long time that I wanted to spend more time with her, and she was right that the quality of time we spend together is different when we are at her house, just the two of us. There, we go for walks, run errands, or go through her clothes like I'm shopping in a fancy thrift store, all the while talking and connecting, with relatively few interruptions. But I also wanted to be seen as a hardworking lab director, professor, and administrator, and amid the flurry of emails and deadlines, it felt hard to say to someone expecting a report or feedback by the end of the day that I wouldn't be able to do it because I needed to hang out with my grandma.

Even if my best self wanted to hang with Bev, my value system was also heavily weighing other immediate demands along with my identity and the opinions of those around me—maybe even more than I'd want it to if I took a step back and more actively reflected on which goals were most important to me in that moment. This is because the value system doesn't operate in isolation, measuring objective rewards and making the same choices no matter what. Instead, it interacts with other brain systems, including ones that deal in who we think we are (the *self-relevance* system) and what we think others think and feel (the *social relevance* system). These were hard at work when I prioritized other things over Bev. I understood myself as a hardworking leader in the lab I had founded, and I understood those around me as people who also prioritized work, maybe parenting, or even being up on the latest trash TV—but not hanging out with their grandmas. These brain systems were foregrounding that information in my value calculation as I considered what my options were for visiting Bev and how important it should be to me.

But Bev *is* important to me, and after her wake-up call, I wanted to change for her. Once I had clarity about that goal, I knew I needed to take a different approach. My research told me that the most salient inputs into my value system were giving me answers day-to-day that weren't aligned with how I wanted to behave. I also knew that one way to change what you think is to change what you think about. I had to find an opportunity to see the situation differently—to help my value system reach the conclusion that visiting Bev is the decision that most resonates with who I am and what I want.

Sometimes it begins with stepping back, noticing what inputs to the value calculation we are prioritizing, and asking, *where are the other possibilities?* Then, sometimes, we see something we hadn't seen before, or a new voice changes the way we understand what was there in the first place. I started looking for a new intervention point, an unnoticed lever to pull.

For me it came from an unexpected source: the podcast *How to Save a Planet*, in an episode by Kendra Pierre-Louis encouraging people to ride their bikes more and capturing the joy that riding could bring to their lives. It's not that I had never ridden my bike in Philly before, but when I thought of riding in the city, I imagined speeding along the way bike messengers do and ending up sweaty and stressed weaving through traffic. Now, as I listened to people on the podcast teetering along on bicycles, laughing gleefully as they gained speed, I started to wonder if this was the lever I had been looking for. If I went at my own pace and used the bike lanes, not only could biking circumvent the traffic and logistical hassles of getting to Bev's house, it could make the journey itself fun.

On a bright fall day, the sun warm on my skin, I stood halfway up on the pedals as I glided down the sidewalk from my house to the corner. I accelerated to the recently repaved asphalt of a bike lane on Spruce Street, past the turrets of the frat houses before the smoother section of bike lane gave way to potholes, and bounced past the hospital complex, on toward the Schuylkill River. On the car-free path, light gleamed off the water, joggers passed people walking their dogs, and I passed the joggers. On my bike I could go fast, faster than running. It felt so free, as though the city—and everything it might have to offer— was available to me in a completely different way. And it was *fun*.

When I got to my grandmother's, we went for a walk, picked up what she needed at the drugstore, continued up her favorite winding residential street in the neighborhood, and looped around to say hello to the statue of General Pulaski behind the Philadelphia Art Museum (she thinks he's very handsome).

Doing it once made it easier to imagine doing it again; this visit gave way to more. Biking to Bev's helped me to feel good about a choice that I had realized was the right one for me—it tipped the scales of my value calculation by moving the "getting there" part of visiting Bev from the aggravating side of the equation to the joyful side, which let me focus on the rest of what I love about those visits. I help her do tasks around her house, we go for walks, and I hear stories about her childhood, about raising my mom, about what it's like getting older. And that feeling of impossible effort? It doesn't feel as hard when I focus on what actually matters most to me, along with the joy of coasting on my bike, the chance to have fun with her, how I never regret having gone.

I still get that feeling of constriction at work when deadlines pile up, or with friends when I realize it has been years since we've meaningfully caught up, but these moments of self-clarity and corresponding change can open space, a crack for light to peek through, a possibility that wasn't there before. It starts with getting curious about why we do what we do, and then gathering possibilities to change. It can mean trying something new, even if you're worried you won't do it right, or listening to the perspective of someone very unlike you. Maybe this will allow other possibilities to take root, grow, and push the crack open a little wider, exploring, reaching out for a new way forward. Maybe you'll be able to see more as the tiny crack widens—and maybe not just for yourself but for those around you. It could mean encouraging your kids to try something that seems scary to them, or helping a colleague say no to adding something else to their overly packed schedule. These kinds of changes can seem small at first, but sometimes these choices mean a lot. After all, you make yourself with what you choose.

So how do we expand our possibilities for choosing? This book explores some key brain systems that shape both what we choose and why we choose. Once we understand why we do what we do, we can explore how we might more deliberately align our daily decisions with our bigger-picture goals and values. In the first part of this book, we'll explore the basic workings of the value system and the value calculation and how we can begin to influence that process. We'll look at different ways of taking a step back to reflect on what is important to us, and we'll see how the way the brain naturally weights inputs to the value calculation may be aligned or misaligned with our bigger goals. In my team's research, we've found that this kind of increased self-clarity translates into greater well-being and a stronger sense of purpose. Understanding the value system helps us see why we make certain choices and can make us more forgiving of ourselves when we regret our decisions or more understanding when others make choices we don't agree with. This lays the groundwork for change.

In the second part of this book, we'll explore how we can change our own behaviors. We'll learn how the brain understands "Future You" much like a whole different person, helping us to understand why it can be so hard to convince ourselves to change by focusing on that future self—as when we try to motivate ourselves to exercise by thinking about how it'll help us live longer, for instance, or go to that networking event with long-term benefits to our career in mind. We'll look at how to turn this insight (and others) into tools for having more agency and bringing our daily decisions into alignment with our goals. With these tools, we can illuminate ways to find more joy and reward in the moment, as biking to Bev's did for me, and in doing so, work *with* the value system. We'll also see how defensiveness can get in the way of transformation. We'll inspect the self-relevance system, which provides inputs to the value calculation, and with this knowledge learn some techniques for becoming more open to—even seeking out—new perspectives, feedback, and change.

The third part of the book is where we'll broaden our lens to see how larger webs of influence interact with our social relevance and value systems to help us change or encourage us to stay the same—and how we can cultivate those influences a little more intentionally. There we'll see what happens in our brains when we communicate and connect effectively and when we don't. We'll delve into neuro-imaging, highlighting how one person's brain can come into synchrony with another's, helping us to connect and communicate. In fact, in a classroom setting, the more students' brains aligned with their teacher's, the more they learned. Likewise, teammates whose brain activity comes into sync perform better on certain kinds of problem-solving tasks. But we don't *always* want to be in sync; there are also benefits to divergence. People enjoy wide-ranging conversations more, and strangers working together on a complex problem strike better deals from exploring new ground. Understanding the value system's role in how we do and don't come together might help us forge the kinds of connections that can lead to our strongest routes of influence. It might help us come closer to the kind of role models we want to be and help us collaborate across differences to create the culture that we want to be a part of.

I hope that if you understand how your brain makes decisions, you might see more possibilities in how to create value for yourself and others. If you're feeling the constriction of cave walls, maybe it will help you to shine a light in a different direction, illuminating the levers that reveal new pathways, the pulley that opens a skylight. This might mean making a change in your own life, seeing a new way through the eyes of someone you admire, or working with others in your community to start a conversation about changes that no one person can make happen alone.

While on the surface this book is about how individuals make choices based on the brain's value system, what I really learned during this research—the bigger, bolder thing I walked away with—is that we have the capacity to consider a much broader range of choices than we think; that we never make a choice in isolation; and that we make ourselves, and the world we live in, with each choice we make.

So how can we embrace that capacity?

It will begin for us as it began for Ayana Elizabeth Johnson, the co-producer of *How to Save a Planet*—the podcast that helped me see a new path forward when I was stuck. Though she has made a career of helping people interact differently with the environment, Johnson's love of the natural world began when she was just eight years old. She was in a glass-bottom boat, peering down at the clusters of different brightly colored fish moving through the coral reef, able to see the ocean from a whole new angle. Sometimes a new perspective can change a whole trajectory, a life.

For us, it won't be the ocean but another extraordinary and mysterious place: the mind. What's really in there? What's the value system up to? And how can we find a flashlight and begin to shine it around, looking for new answers to choices big and small?

# A NOTE ON THE RESEARCH

To study the brain, my lab primarily uses functional magnetic resonance imaging (fMRI), which measures changes in blood flow in the brain as a proxy for neural activity. Since all the cells in your body need oxygen to work, and blood brings them fresh oxygen when they need energy, an fMRI scan allows us to get an idea of where neurons are firing most heavily (where the most blood is flowing). Through fMRI scanning, researchers can get a sense of how and where brain activity *changes* when people are presented with different stimuli, including visuals shown on a computer screen; audio played through headphones; and various tasks they can engage in by pressing buttons, using a joystick, or following along with their imagination.

Scientists use this technology to observe what is happening throughout the brain without necessarily having to interrupt people to ask what is driving their thoughts. This is significant, because such questions could change the very processes we are trying to observe: How emotional is this decision? Is your process automatic or more effortful? Are you relying on social thinking, emotions, basic sensory inputs, memory? How much of this is driven by your identity? Measuring brain activity gives scientists information that complements people's reports of their perceptions, preferences, and intentions, which helps us understand and predict their future choices.

In this book, we will explore a wide range of neuroimaging experiments that use fMRI and that share these benefits but also some core limitations. In some areas of cognitive neuroscience, we know a lot about the function of different brain regions or networks of brain regions, so when we see those regions activate, we have a good idea about the types of thoughts or feelings people might be having. For example, seeing activity in the brain's visual cortex, scientists can pretty impressively reconstruct the type of image the person was looking at. But when we move beyond lower-level sensory experiences, things become more complicated. When it comes to higher-order thoughts about our own identities or how we make sense of other people and situations, brain scans can't reveal exactly what any individual is thinking. In the best cases, neuroscientists like me are making educated guesses. For example, we might infer from the activation of certain brain regions that people are experiencing a sense of reward or are thinking about their own or other people's thoughts, but we wouldn't know for certain, because most brain regions do multiple things. This means that we couldn't specifically see what thoughts they imagined that other person to be having or specifically whom they had in mind, since each brain region serves so many different functions.

Another major limitation of most of the neuroimaging research that I'll share in this book is that the research participants whose brains were scanned represent only a very small sliver of humanity. Functional MRI requires costly equipment that is typically operated at major research universities, and each brain scan is expensive. It is also convenient to study college students as participants. In part for these reasons, many of the early studies in this field were limited to white, Western, educated young adults; information about other important dimensions of participants' identities, like their religion and sexual orientation, often isn't measured or reported. In addition, the results we'll explore in this book come from averaging across many people's brains; since each of our brains works a bit differently, these findings represent some of what is common across the groups of people scanned, rather than what is true for everyone. Although more recent work is actively trying to address this major gap in our knowledge, there's a ton we don't yet know about whether and how specific conclusions might change for different people, holding different identities, across different cultures and contexts.

Finally, this field is very new relative to other social and biological sciences. We are learning more and more about how human brains work, and how they vary between people and across time. This makes it especially exciting to do this work now, but also means there is still a lot to learn. This book is, in a way, a snapshot of what we understand now. I expect that, as with science itself, this understanding will grow and evolve over time. It is a powerful time to stand on this frontier and look toward the horizon.

# PART 1

# CHOICE

# 1

# The Value Calculation

JENNY RADCLIFFE IS KNOWN online as "The People Hacker." There are many ways she describes her job: a "burglar for hire," a "professional con artist," a "social engineer." But officially, she's a "penetration tester"—a security consultant whom companies hire to break into their buildings and computer systems to help identify weaknesses in their security infrastructure.

Although Jenny sometimes uses physical force, lock picks, or computer code, her main tools come from psychology. She can read a person or a situation and predict how someone (or a group of people) will respond to her, depending on what she does. Then she can create a situation that moves her toward particular goals and outcomes.

This is just what she did when she was hired to break into a bank in Germany. Her mission was to enter the bank during business hours, get past security, and locate a particular office, where she was to plug a USB drive the company had given her into a computer. A program preloaded on the drive would then install itself on the computer, letting the company know that Jenny had successfully penetrated their security.

The morning of the big job, Jenny readied a costume and props.

She wrapped her hand and wrist in a bandage, figuring that people might be more likely to hold doors open for her if she appeared to be injured. She brought a big file box full of papers to occupy her hands, further increasing the odds people might hold doors for her. Thus prepared, she went to the bank, walked into a grand lobby furnished with leather sofas, and approached huge doors blocking access to the "employees only" portion of the bank.

Those doors presented Jenny's first of many obstacles. They were operated by fingerprint scanners, and of course, Jenny's fingerprint wasn't in the bank's system—she wasn't an employee, she was pretending to be one. But she walked over to the fingerprint scanner and put her finger on the pad anyway. It beeped—no luck. She hadn't expected the sensors to let her in, but as a penetration tester performing a security audit, it was still part of her job to check.

At this point, Jenny had choices. She could ask the security guard on duty in the lobby to let her in, but what incentive would he have to do that? It was his job to keep strangers out. So instead she did the obvious thing: she swore, really, really loudly.

Just as Jenny had planned, the security guard came over to see what was happening.

"You don't have to work on the lock," Jenny later explained. "Work on the person behind the security. It doesn't matter what they put in place; if someone's got access, then I can access them, and then we're down to me versus the person."

When the guard approached, Jenny said impatiently, "This isn't working. It was working yesterday." The security guard suggested that she try the fingerprint sensor again. She made a big show of being annoyed, cursing once more and awkwardly balancing her big box of papers on her bandaged hand. She tried again; the machine beeped again. Maybe she wasn't pressing hard enough, the guard ventured. She grudgingly placed her finger on the sensor again—at which point the guard took her hand and tried to help her press her finger onto the machine.

Jenny yelped in apparent pain and swore loudly once again. She made a point of dropping the file box, which scattered papers everywhere, and made a big show of trying to pick them up, all while swearing away. Now she had drawn attention to herself—people in the lobby were looking.

"For God's sake, go in," the guard said, and beeped her through the doors. "Thank you, danke schön," Jenny replied. And she was on her way—down the hallway to the designated office, where she inserted the USB key she had been given.

What happened here? Making a big commotion like Jenny did might not work for every person in every situation. For one thing, some people might be more influenced by being buttered up or feeling like they're doing someone a favor. For another, the same actions can be interpreted as more or less threatening, depending on the characteristics of the person doing them and the environment they are in. But in this case, Jenny felt confident that causing a scene would help her break into the bank because she knew that in Germany people generally feel highly embarrassed by a scene, and based on her gender and the way she looks, she wasn't likely to be perceived as a physical threat or a computer hacker. Under these conditions, making the commotion the most prominent thing in the guard's mind would tip the scales of his decision-making. She figured that the guard would perceive her as low-risk and would rather buzz her in than deal with the discomfort and disturbance of a spectacle. And she was right.

Maybe you feel tempted to harshly judge the guard for letting Jenny in. The bank's rules no doubt emphasized that he should not let strangers through the door. If Jenny had been a malicious hacker, the USB drive she plugged in could have uploaded a computer virus that stole customers' personal information and life savings or taken down important parts of the bank's infrastructure. But the truth is that many of us would do the same thing in that situation. We want to see ourselves as helpful, kind people, and much of the time other people aren't trying to deceive us. If Jenny *had* been an injured employee simply trying to get into her office to do her job, the guard's actions would have been helpful to the bank, not harmful.

For better or worse, Jenny's understanding of these decision-making mechanics—the sometimes-unconscious, near-instant calculus we perform when choosing between options—and how they can be influenced enabled her to break into the bank. Recent advances in neuroscience allow us to understand more about the underlying systems in the brain that allowed her to do this, and that might allow others to resist, including one that scientists call the value system.

As we begin to explore the value system, which brings together many different types of information to guide our decisions, it may be helpful to imagine the thought process of the security guard when he was confronted with Jenny. His brain's value system would compute the value of different possible choices (allow the swearing woman to continue making a scene or buzz her in), select the one with the highest value (buzz Jenny in), then track how rewarding the choice is (now the scene is quiet, and I feel good that I helped an injured person). Much of the time, this value calculation happens quickly and seamlessly. Importantly—as Jenny understood so well—its outcome depends on what our brains pay attention to in the moment. In that split second, the value calculation can be shaped by any number of factors: our own goals, how we feel, our identities, what we think others will think and feel, other people's actions, cultural norms and expectations, our social status, and much more.

Jenny used her implicit understanding of the value calculation to gain access to the bank, as she had been hired to do. Now alert to this vulnerability, the bank, in turn, might take steps to ensure a different outcome to guards' value calculations in similar situations in the future. Making the guards aware of how Jenny broke in could empower them to exert more agency over their decision-making in such a moment and resist future attempts to hijack it in that way. Or, the bank might provide more opportunities for security guards to get to know the other bank employees so that it would be clear when a new employee joined, as well as who was a stranger.

Of course, to think of all these options requires thinking along a number of different dimensions: checking in with the bank's big-picture goals, the security guard's goals, and where there might be room for greater possibility in the overlap. So what options, or combinations of options, would make it more likely that the security guard chooses differently next time? How might we become more aware of when our value calculations are being shaped by people who don't have our best interests at heart? To figure this out, it's helpful to know what's going on in our brains when we are confronted with choices.

## KOOL-AID OR PEPPERMINT TEA?

One remarkable power of the value system is that it allows our brains to take complicated, messy, real-world decisions and boil them down into comparable quantities. Thus simplified, our brains are able to choose between options—often almost instantaneously and with a fair amount of internal consistency.

I find it useful to think of the value calculation as a hidden game of "Would You Rather?" You're probably familiar with this common icebreaker, in which one player offers two (ideally silly) choices, and other players say which they would prefer: Would you rather have a cat's tongue or roller skates for hands? Would you rather be able to speak every language or have the most beautiful singing voice on earth? Would you rather live alone on a desert island with all the movies and books ever made or with one other person you choose, but no media?

When you think about it, it is borderline magical that you *can* answer "would you rather?" questions, comparing alternatives that differ in so many ways. From low-stakes situations like playing the game "Would You Rather?" at a party, to the decisions that determine our actual behavior each day, our value systems help guide us to our choices. But *how* does the brain do this?

For a long time, no one knew the answer. Did the brain have different systems that each monitored different dimensions of a choice? (How much sugar or salt is in each food we are choosing between? How hot or cold is each food? How green is each food?) Or were there different brain systems that would handle choices in different domains? (A brain system that decides what kinds of foods we want to eat, a different brain system that keeps track of how much fun each of our potential dinner companions is, and a third that handles the financial decision about whether we can afford to eat out?)

The foundations of how we currently think about the neural underpinnings of this kind of decision-making were laid in the 1950s by researchers who mapped a set of brain regions that tracked simpler types of rewards and that guided animals' behavior to maximize those rewards—even if choosing the reward was objectively bad for the animal's well-being in the longer term.

James Olds and Peter Milner, scientists at McGill University in Canada, discovered that when given the chance, rats repeatedly pressed a lever that triggered electrodes that stimulated particular parts of their tiny rat brains that made them feel good. In other words, the rats found it "rewarding" to stimulate these parts of their brains, and scientists at the time began to think of the regions being stimulated as the "reward system." It turned out that stimulating this reward system had powerful consequences for the rats' behavior. For example, when rats were given the chance to press a lever that stimulated these reward regions, they would even forgo food that they needed to stay alive.

And it wasn't just rats. Scientists soon found parallel reward systems in rhesus monkeys and eventually came to learn that all mammals had similar infrastructure in their brains. Across species, when scientists stimulated neurons (the cells that transmit messages through the nervous system) deep in the brain in a region called the striatum and in certain regions in the front of the brain (frontal cortex), the animals seemed to experience reward, as evinced by their tendency to seek out the stimulus over and over. Like humans, some animals also displayed facial expressions or made sounds showing their pleasure. But although it was clear early on that stimulating specific reward regions caused animals to want things, it took several decades for scientists to understand how this translated into more complex decision-making in humans. Why would a system that tracks how much food you want or how much you want to press a lever have anything to do with whom you want to be president or which movie you want to see? Could a single brain system really handle comparing choices that take place at various points in time (now versus later), concrete rewards like which snack to eat, and abstract questions about society and morality?

A series of important insights about how brain systems make more complicated calculations about the relative values of a wider range of goods and ideas came in the mid-2000s—one of them through offering Kool-Aid to monkeys. Camillo Padoa-Schioppa and John Assad were researchers at Harvard Medical School studying decision-making and economic choices when they wondered whether the reward system discovered in rats and other animals could also help monkeys make somewhat more complicated decisions, and if so, how? On the one hand, they reasoned, it was possible that regions of the reward system might respond to objective properties of different potential rewards (like the amount of sugar in a juice). This might be the case if a particular nutrient, like sugar or fiber, had been important to the survival of the species in the evolutionary past, and a physical feature of the food, like color or firmness, was a good indicator of how much of this nutrient was present in it. If so, there should be a tight correspondence between certain biological or chemical properties of foods and the response of the reward system. On the other hand, what if the reward system could take a wider range of things into account, to make more subjective calculations? Could it explain why a monkey might have different food preferences at different times—or even predict what a monkey was in the mood for?

In their experiments, Camillo and John would present a monkey— let's call him Gizmo—with a series of choices while recording the activity from neurons in his brain. Would Gizmo like one drop of lemon Kool-Aid or two drops of peppermint tea? Five drops of milk or one drop of grape juice? Gizmo would look left or right to indicate his decision.

After many of these choices, the researchers could calculate how much value Gizmo assigned to each drink relative to the other drinks—what neuroscientists now call its *subjective value*. We say the value is subjective because it turned out not to be fixed to some objective quality like the density or overall amount of sugar present in each liquid, the exact temperature, the quantity of liquid, and so on. The scientists found that Gizmo and other monkeys generally preferred to have more to drink, if possible, but, like humans, they liked some drinks (specifically, lemon Kool-Aid and grape juice) more than others. Depending on the offer, the monkeys would sometimes choose a smaller amount of their preferred drink over more of one they liked less. By offering the monkeys the drinks in different ratios, Camillo and John could arrive at a mathematical description of the monkeys' preferences in each session. For example, if Gizmo was really in the mood for grape juice in one session and chose one drop of it over up to three drops of water, then Camillo and John could say that one drop of grape juice was worth three points, while one drop of water was worth one.

While hanging out with the monkeys, Camillo and John also found that subjective value was influenced by the *context* within which the decisions were made: the monkeys' drink preferences (that is, the relative value of one drink to another) varied from day to day—even for the same monkey. Imagine that you yourself are at someone's house and they offer you a cup of coffee or a cup of lemon-ginger herbal tea. Your decision depends partly on stable preferences you have (you typically like coffee more than lemon-ginger tea), but also on the situation (it's late and you worry that caffeine might make it hard to sleep). Similarly, on Tuesday Gizmo might prefer grape juice to water 3:1, but on Friday he might feel less strongly because he's already had plenty of fruit and may prefer the grape juice to the water only 2:1. This is what "subjective value" means—different aspects of a situation change how much something is worth to someone, at a given time, in a given situation.

When Camillo and John looked at the data from the monkeys' brains, they discovered that neurons in the front and center—specifically, a region called the orbitofrontal cortex—fired in response to each monkey's overall *subjective* preferences for the juices. The activity in these neurons correlated with the overall ratios Camillo and John had calculated based on the monkey's decisions—when the monkey preferred one option three times as much, these neurons fired correspondingly more. Interestingly, the firing didn't seem to depend on objective aspects of the choice, such as the specific ingredients of the drink (if, as you might think, there were neurons tracking the amount of sugar), which side of the screen showed the offer (if neurons here kept track of what motion the monkey needed to perform to get juice), or how many drops of juice were offered in total (if more is always better). Instead, the neurons tracked the overall, *subjective* value.*

And this subjective value was tied to the choices the monkeys made. Just by seeing what was happening within Gizmo's orbitofrontal cortex when he was shown the different options, Camillo and John could

---

* Neuroscientists sometimes use the words "reward" and "value" interchangeably, though "value" commonly refers to expected outcomes, and "reward" the actual outcome. Other researchers think of the process of calculating value as a more deliberate, cognitive process, whereas the experience of reward is more basic and pleasure-based. In this book, we will often use these words interchangeably, though the basic idea that animals have a reward system has been around for much longer than the recent insights about how the brain makes more complex calculations related to what we value.

predict which choice Gizmo might make with remarkable accuracy. In other words, the monkeys' brains were computing subjective values for each option on a common scale that allowed them to make decisions and compare apple juice and orange juice.

But what about humans? Around the same time that studies on monkeys revealed that their brains responded to subjective (rather than objective) value, scientists began to find similar responses in the human brain. In the span of a decade or so in the early 2000s, scientists ran hundreds of experiments mapping what happened in people's brains when they made choices based on these subjective preferences.

In one early study, the neuroscientist Hilke Plassmann and her colleagues at Caltech found that when they measured how much human volunteers were willing to pay to eat different snacks, they showed similar activity in brain regions analogous to those the monkeys used to choose between lemonade and grape juice. The team showed pictures of salty and sweet junk foods, like chips and candy bars, to hungry humans while scanning their brains using functional magnetic resonance imaging (fMRI). This type of brain scan lets scientists see when different parts of the brain are active and then connect this activation to different psychological processes and behaviors. The volunteers in Hilke's study were told they had a specific budget and were asked how much they would be willing to pay for different food items, shown as images on a screen in the fMRI scanner.* As in the case of Camillo and John's monkeys, brain activity increased the most within a similar region in humans—the ventromedial† prefrontal cortex—

---

* Importantly, once the volunteers got out of the scanner, one of the snacks was randomly selected, and if the price they had been willing to pay was equal to or lower than the actual price of the food, they would receive the food, along with change from their budget. If the actual price of the food was higher, they would simply receive the total budget in cash as payment at the end. This incentivized participants to report a price they were truly willing to pay for each product, revealing their honest preferences and therefore encouraging them to behave just as they would outside the lab.

† The ventromedial prefrontal cortex is the lower portion of the broader medial prefrontal cortex.

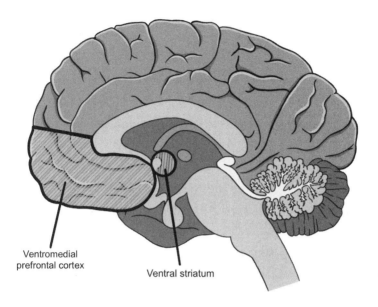

*The ventral striatum and ventromedial prefrontal cortex, pictured here, are key regions in a broader system that tracks subjective value when people make decisions across many domains.*

for the items they rated as most valuable. In other words, there was more activity in response to snacks they were willing to pay $3 for than snacks they were willing to pay $1 for or didn't want to buy at all. People's brains kept track of the subjective value (to them, personally) of different foods, and chose accordingly.

This was a breakthrough—but in daily life, we often have to choose between options that are harder to compare than two kinds of snack foods. Could the same brain regions that decide if you'd rather drink coffee or tea also compare things that are rewarding in very different ways—for example, would you rather drink grape juice or go see a movie?—or did such choices go beyond their role in decision-making?

To probe this question, a team of scientists at Caltech and Trinity College Dublin designed an experiment that was, in essence, a variant of the "Would You Rather" dilemma: The research team gave volunteers in an fMRI scanner a $12 budget that they could use to bid on different types of goods, from sweet and salty snacks, to DVDs, Caltech memorabilia, and monetary gambles. They found that an overlapping area of the ventromedial prefrontal cortex tracked how much people were willing to pay not only for different foods but also for products like college memorabilia and DVDs. Around the same time, other groups of scientists were also finding that activity in the human medial prefrontal cortex and other regions, like the ventral striatum, tracked people's willingness to pay different prices for a range of consumer goods. These findings suggested that a common system was keeping track of the value of a wide range of different kinds of choices.

As this body of research grew, this group of brain regions, including the ventral striatum and ventromedial prefrontal cortex, came to be known as the value system. By 2010, activity in the value system had been shown to track not only people's decisions about how much money they would pay for different goods, but other kinds of financial choices as well. For example, would you prefer to take a 100 percent chance of winning $10 or a 50 percent chance of winning $20? Would you rather have $10 now or $20 in six months? All these types of choices seemed to work through a similar mechanism in which the value system identified and assessed the subjective value of different choices, compared them, and then acted.

By 2011, researchers could even predict, based on activity observed in volunteers' value systems while they were looking at different goods, what they would later choose—even when they weren't asked to make any choices during the initial scan. In other words, the value system seems to track the subjective value of different things regardless of whether the person is consciously trying to make a decision about them. When we're in line at the grocery store, our value systems are weighing the value of the candy bars by the register and absorbing information from the news headlines and magazine covers. When we're scrolling through social media, passively consuming ads, our value systems are still registering the inputs, even if we aren't actively paying attention to them.

A decade later, it is now more widely accepted that our brains can make calculations using a "common value" scale that allows us to compare things that aren't inherently comparable. You could probably easily decide if you'd rather snuggle a puppy or have $5 right now. This is because your value system converts each option onto a common scale and makes the comparison. Likewise, when Jenny yelled for the security guard, he quickly made the decision to try to help her use the fingerprint scanner, rather than demand ID, and eventually to let her through the doors, rather than calling for backup, asking her to leave, or asking her on a date.

## PREDICTING AND LEARNING

It's tempting to think there are *good* choices and *bad* choices, but the truth is that these are moving targets, and the value system is dynamic, constantly weighing competing interests and the context. This means that the choices we make depend on what options we imagine we are choosing between and what dimensions of the choice we focus on. If your kid has never met a male nurse, it might constrain the career options he imagines choosing to suit his empathic personality. Moreover, the subjective value we assign to a given choice option can change, depending on a variety of factors related to our past experiences, our current situation, and our future goals. If your kid believes you'd like him to get a job that helps a lot of people, that dimension might weigh heavily as he considers career options. Likewise, if his crush gushes about Austin, Texas, that might cause your son to give weight to the geographical flexibility of different job options. This is one neural foundation of what social psychologists call "the power of the situation": our decisions depend on our current context, which gives certain inputs to the calculation more weight.

Let's say you're deciding whether you'd rather eat a salad or chocolate cake. If your brain only followed "objective" rules, you might only care about how much the food filled your stomach or how many calories it offered (which could translate directly to keeping you alive in earlier moments of human evolution). But that's not how it works. As you have no doubt experienced, when you decide what to eat, you might focus on any number of things: how does the food taste, how will you feel after eating it, what is your date eating, did you just get a bad doctor's report, do you have a great metabolism, is it someone's birthday, how much does each cost, did you just run a marathon, are you in a bad mood? Your brain does this quickly and may not even take into consideration all these dimensions, limiting what it weighs in any given choice. Based on what factors it does weigh, your brain can compute subjective values for salad and cake on a common scale, then choose the higher-value alternative.

Once you've made the choice, your value system transmits it to the parts of your brain that help you act on the decision, like reaching out and grabbing your chosen food and eating it. Importantly, your brain's value system then keeps track of how good the decision's outcome was, relative to what you thought would happen—in other words, how accurately it guessed how rewarding the choice would be. It tracks not only your prediction (That cake looks delicious! I remember how much fun I had at birthday parties as a kid!), but the *prediction error*, or the discrepancy between your prediction and the actual outcome. If the choice ends up being more rewarding than you expected (That cake was delicious! Totally worth it!), your brain generates what neuroscientists call a "positive prediction error," seen as an increase in activation within the value system after the choice; conversely, if the choice ends up being worse than you thought (That cake made me feel gross!), your brain generates a "negative prediction error," seen as a decrease in activation within the value system after the choice. These prediction errors help you learn for the future, updating how your brain makes the value calculation over time.

In sum, there are three basic stages to what neuroscientists call *value-based decision-making.* First, our brains determine what options they are choosing between, assign a subjective value to each one, and identify the option with the highest value *in that moment.* This means that from the start, our choices are shaped by what we consider the possible options in the first place. Next, our brains move forward with what is perceived as the highest-value choice (which may or may not be the *best* choice in the context of our larger goals or longer-term well-being). This means that there isn't one single right answer, and what our brains perceive to be the "highest-value" option right now might change if considered from other perspectives (for example, when thinking about career goals versus wanting to be a good friend). Finally, when we've made the choice, our brains track how rewarding it turns out to be, so they can update how they make the calculation next time; this means that we often overweight the outcomes of our choices rather than improving our process. This highlights at least three places where we can intervene: we can imagine more (or different) possibilities; consider the existing possibilities from different angles; or pay attention to different aspects of the outcome.

We can think again of our security guard. If, as the guard, you buzz in a bumbling person making a scene and it yields a better social reward than you had expected (the person gives you a big, grateful smile and tells you how much she appreciates you), your brain will generate a positive prediction error, that data will be stored, and in the future, you will be more likely to let in the next bumbling stranger. But if something bad happens and the outcome is worse than you anticipated (the bumbling person turns out to be a security tester and your colleagues are annoyed with you because now you all have to sit through extra training sessions), your value system stores that too. Next time, you might think twice before letting in a stranger.

But, of course, no one scanned the brain of the security guard. Most of the studies we've explored so far have taken place in highly controlled lab settings. So what actually happens outside the lab, in the real world? Can we link activity in the value system to what people do in their day-to-day lives outside the brain scanner?

## A GREAT DAY FOR SCIENCE

I was a budding neuroscientist in the early 2000s, when our understanding of the value system first started to take shape, and I was interested in whether brain imaging could give us insight into health decision-making. I wanted to help people make choices that would help them live healthier, happier lives, but I also knew that these choices could be very difficult to make. It's hard to change, and even when we *are* motivated to change, we don't always take time to figure out why we do what we do in the first place or know why some ways of thinking are helpful in achieving our goals, and some aren't.

I was thinking about how to make better health coaching and messaging campaigns. I was also thinking about how we might talk with our family members and friends, roommates and colleagues, to help motivate them to make healthy changes, and even how we might talk to ourselves to make decisions that are more in line with our goals. I wondered if brain imaging could give us a new window into this decision-making. Maybe looking at brain responses to health campaigns and health coaching messages could help us understand what made people change and what would make it easier to work with, rather than against, our desires. If that were true, maybe it could help us design and select better messaging.

I decided to apply to graduate school to work with Matt Lieberman at UCLA. Matt's lab was full of scientists studying how people understood themselves and others and how they made important decisions. Along with a group of other young faculty, Matt had recently ignited a new field of study that combined social psychology with cognitive neuroscience; whereas neuroscientists before had focused on topics ranging from vision and memory to reward and motor actions, many fewer had delved into topics that were more at the core of being human, like where our sense of self comes from, how we understand what others think and feel, and how imagination works.

At the time, it felt like a long shot to connect what happened in a neuroimaging lab to real-world behavior changes outside the lab. But it also felt fundamental: what good was all this research if it couldn't help us in real life? Luckily, during the years I was in graduate school, we *did* start to see a connection: a pattern indicating that activity in the brain's value system could reveal who is more likely to change their behaviors in response to messaging and what kinds of messages were most likely to elicit this kind of activity.

The first work we did in this space focused on sunscreen use. In Los Angeles, where it is sunny almost every day, I had a daily reminder that—despite how great the sun feels warming your skin—sunburns and other invisible damage from UV rays can cause skin cancer. Matt and I designed an fMRI experiment, scanning the brains of volunteers while exposing them to messages about the importance of wearing sunscreen every day.

The finding was simple: the more activation we saw in a person's value system—specifically, the ventromedial prefrontal cortex—in response to the messages, the more likely they were to increase their sunscreen use in the next week. It suggested that the value system helps guide not only simple choices that people make in the lab but also real-world, consequential behavior change outside the lab.

When I saw the data, I started jumping up and down on the lab couch. My friend and then-officemate Sylvia claims that I screamed, "This is a great day for science!"* While I don't know if nonscientists would be this excited about a data plot, it felt like a big moment. And although this initial study relied on what people told us about their sunscreen use, later studies in the lab I now run at the University of Pennsylvania and others have shown similar results in people being

---

* I deny it.

coached on other health habits, where behavior change has been measured more objectively.

When sedentary adults were exposed to messages encouraging them to get more exercise, the activity in their value system corresponded with how much exercise they got later, measured objectively using wrist-worn activity trackers. Similarly, smokers whose value systems responded more strongly to messages encouraging them to quit smoking were significantly more likely to reduce their smoking over the following month, which we confirmed using a device that measures how much carbon monoxide smokers have in their lungs. In fact, our ability to predict how much people would reduce their smoking was twice as good when we included information from both brain responses and self-report surveys as when we included only information from the surveys. This suggests that there was useful information that the value system captured that was not fully captured by surveys alone. Figuring out why this is the case, and how far in the future we can predict, is a current frontier.

Another current frontier involves understanding when and how people make the kind of deliberate decisions that we'll mostly focus on in this book, compared with other kinds of decisions. For example, it is increasingly clear that a lot of what humans do is guided by habitual routines—which is not the kind of choice we'll be discussing. But some of these habits start with deliberate choices, which *is* our focus. To illustrate this distinction, let's consider my walk to work.

When I first moved to Philadelphia, I wanted to walk to work, rather than drive or take the subway, so I'd get outside more—that was an active choice. I used my phone's map to find the shortest route, and following my phone's map was also an active choice. Over time, as I repeated this walking route over and over, it became a habit—something I could do (and did) on autopilot, whereas other options like driving, taking the trolley, or even walking a different route require more conscious thought. In other words, when repeated over and over, what start as goal-directed, value-based decisions become routine and get handed over to another brain system that supports the kind of automatic pilot I was on. This book explores what happens in the first type of decisions—when we are more deliberately choosing and setting in motion paths that may (or may not) eventually become habits.

## CHARTING A NEW PATH

My partner, Brett, and I don't usually walk to work together, but one morning the stars aligned to make one of these mini-dates possible. Then, as we set out, Brett turned down Osage instead of walking up to Pine. *This isn't the right way to campus*, I thought, annoyed. But as we walked, he pointed out beautiful buildings, interesting turrets, arches, and other charming details he likes about the fraternity houses that line that street. He had discovered them by investigating different routes to work every morning. Instead of going on autopilot like I typically did, he had decided to take advantage of his morning walk to work as a time to see new things, a series of small adventures that enriched his day. The various paths from our house to the University of Pennsylvania are all pretty much the same length, so it wouldn't even cost me time to try something new. It might even lead me to discover more of the world around me, have more interesting things to share with people I care about, and just generally be the kind of person who looks for little adventures each day. What other opportunities was I missing?

This makes it worthwhile to do an audit every once in a while and to work toward developing an awareness of *why* we do what we do. What are the everyday choices we're making? How are we making them? Are there new choices that we can make or ways to choose differently? Are there possibilities we haven't even considered? And are the choices we are making really serving the lives we want to lead, the people we want to be?

Although this is a bit oversimplified, what we do when we ask these questions is bring into play brain systems that can help probe and shape the value system's workings.

In fact, the value system works in coordination with many other brain systems, including sensory inputs (what am I seeing, hearing, smelling, touching, tasting?), memory systems (what has been rewarding to me in the past?), and attention systems (where is my current focus?). Brain systems involved in reasoning and regulating our emotions can also change how much weight we give to different inputs. For example, as we'll explore more in Chapter 4, I might give more weight to how tasty or how healthy different foods are, depending on my goals. By observing what happens *throughout* the brain when someone is presented with a choice, neuroscientists like me have seen that the value system synthesizes and uses many kinds of information to arrive at a decision.

In the chapters that follow, we will put special focus on two brain systems that influence valuation and have emerged as especially important in decision-making. The first, called the *self-relevance system*, helps us understand ourselves. The self-relevance system is concerned with questions I call "Me or Not Me"—questions like: What do I care about? What has happened to me in the past? What might I do in the future? Although the details vary from person to person and context to context, in general we categorize things in terms of their personal significance to us (whether they are "relevant to me" or not), which in turn shapes how much effort we put into making the choice and also shapes the personal rewards we expect from different choices. The brain then creates the feeling that something is "me" or "not me" in relation to what I like and value. Jenny likely used this brain system to summon her confidence that she's the kind of person who can pull off a stunt. In turn, the guard's self-relevance system may have reminded him of an identity as a helpful person. A helpful person would assist the struggling employee who was having trouble with the fingerprint scanner. Whether a given option feels like "me" or "not me" can influence the outcome of the value calculation, as we'll explore more in Chapter 2.

Another key input to the value calculation comes from the *social relevance system*, which helps us understand what other people think and feel: What do *you* care about? What knowledge do *you* already have? What might *you* do next? This knowledge helps us think through more specific questions, like: Why didn't you answer my text message? Do you like jokes? How will you respond if I hug you? The human brain has evolved to help us make sense of other people and to evaluate what someone else might think and feel. Jenny was using this brain system when she formulated her plan to break into the bank, guessing about how the security guard would react. In turn, the guard's brain was likely leaning on input from his social relevance system when he saw the effects of her commotion and made the decision to buzz her through. Our social relevance system allows us to simulate (sometimes accurately, sometimes not) what happens in other people's minds, and the value system uses this information to guide our own choices. We will explore the social relevance system more in Chapter 3.

As a neuroscientist and as a person, I find this knowledge empowering. Knowing how flexible, dynamic, and influential the value system is—how many different factors it is able to weigh in a given choice—helps me to appreciate my own and others' ability and potential to change, adapt, and grow. Once you understand how the brain assigns value to different options, you can view the decisions you can make with a broader lens. I like to think of it as a way of exploring the question, Where is the possibility? It's a way to direct a flashlight around in the dark, finding crawl spaces, escape routes, and paths forward that you might not have otherwise realized were there. As we've seen already, the value our brains assign to any given option is never fixed. Your behavior isn't determined solely by your genes or your education or your personality, and it is highly dependent on context and culture. Understanding this, someone like Jenny can guess some of the factors that go into a person's value calculation and construct a situation that spotlights those that serve her goals. But it doesn't only help faux bank robbers. By understanding these principles, leaders at the bank might see the situation from the guard's perspective and offer solutions that appeal to the security guard's identity as a helpful person, while also protecting the bank. In the same way, we can influence what our own and others' value calculations focus on and in so doing potentially change the outcome—bringing our daily choices in better alignment with our bigger goals by expanding the range of possible options before us, and noticing where we might be vulnerable to influences that go against our goals and values. But to understand what these possibilities are, we first need to understand who we think *we* are, at our core—and what that has to do with the way we make choices.

# 2

# Who Am I?

IF YOU'VE FOLLOWED AMERICAN popular culture over the last fifteen years or so, chances are good that you've encountered the work of actor and comedian Jenny Slate, whether you know it or not.* You might have seen her capering on *Saturday Night Live,* or stealing scenes as the irreverent, and often crude, Mona-Lisa Saperstein on the beloved sitcom *Parks and Recreation,* or starring in the award-winning romantic comedy *Obvious Child.* You might have watched her stand-up specials on Netflix. Or maybe you've only heard her voice, in kids' movies like *Zootopia* and *Despicable Me 3,* animated series like *Big Mouth* and *Bob's Burgers,* or her Oscar-nominated stop-motion film *Marcel the Shell with Shoes On.*

Whatever character she's taking on, one thing that stands out about Jenny is that, somehow, she always remains strikingly herself. But this wasn't always how she felt. As a cast member on *Saturday Night Live,* Jenny tried to be the kind of attention-grabbing comedian she saw modeled by others. The show is a legend of American culture that generations of comedians have dreamed of joining, and she liked the

---

* I promise that not all the characters in this book will be named Jenny.

people she was working with. But something wasn't quite right. She sensed she wasn't really being herself.

There was a way that she thought comedians were "supposed to" act—a certain gutsy, sassy personality they projected—and she spent a lot of time thinking about how she could present herself in that mold. "How do I seem like I just don't even care?" she'd ask herself. The problem was, that's not who Jenny is. She does care. She cares a lot.

And it wasn't just that the job didn't feel like the right fit—she felt like she'd failed. She felt like the problem was *her.*

Then, one day, she found a voice, a sweet, gravelly, child's voice. She was crammed into a hotel room with friends at a wedding, and the voice emerged as a spontaneous, authentic expression of how small and crowded she was feeling in that hotel room, and in other moments of her life too. It wasn't *her* voice, not exactly—but also it was. And it made her friends laugh.

After the wedding, Jenny and her partner at the time, the director Dean Fleischer Camp, decided to make a character out of it. Dean bought supplies from a local craft store, and they glued the pieces together to create Marcel, a small creature in the body of a seashell sporting one wide, googly eye and little pink shoes. They filmed a short stop-motion video in which Marcel describes his life and what he's like. We learn that he wears a lentil as a hat and hang glides on a Dorito for adventure. "My one regret in life is that I'll never have a dog," Marcel the Shell tells the camera. "But sometimes I tie a hair to a piece of lint and drag it around." Marcel is also unapologetic about who he is. "Sometimes people say that my head is too big for my body and then I say, *compared to what?!*"

Jenny and Dean posted the video, "Marcel the Shell with Shoes On," on YouTube to share with friends and family. Open with his feelings, at moments modest or shy but also straightforward and honest, charming, and deeply sincere, Marcel is much like Jenny, but very different from how you probably imagine the work of a typical American comedian. And so the video's reception stunned her. Instead of only a few friends and family, as she had been expecting, over thirty-six million people watched it. And, much like her friends had, the audience responded with love.

The experience of having people embrace this authentic expression of herself was revelatory. "In that moment," she relates, "it was worth it. . . . Like I knew magic exists."

She realized she didn't have to be someone she isn't, or act like she doesn't care, to connect with an audience. Instead of fighting against or suppressing her core sense of self, she could make art that expressed it. Jenny found a new possibility for her life and work, and it felt autonomous, self-concordant, easeful. The world of comedy suddenly felt wide open.

In the years since, the feeling of who she is has guided the way she thought not only about Marcel (whom she brought into additional shorts, a children's book, and a full-length feature film) but also her stand-up, her book *Little Weirds*, and other projects as well. Listening to the voice that tells her when something doesn't feel right ("this isn't me"), and, as importantly, listening to the one that does feel true to herself ("this is me") has led Jenny to make art that is not only funny, but also connects deeply with her audiences.

This ability of Jenny's to determine and notice what feels like "herself" is supported by the brain's self-relevance system. Neuroscientists have mapped brain regions that people use to construct a sense of what is "me" and "not me" based on our past experiences, our current context, and our future goals. The self-relevance system helps us answer questions about our current mental states and our broader traits, how our life choices and experiences fit together to make a more coherent story of our lives, and how to make choices that will be rewarding to us.

If this sounds a lot like a value calculation to you, it's not a coincidence. The self-relevance system and the value system are highly intertwined within the brain, making self-relevance a major input to the value calculation, and vice versa. This means that when we're faced with a decision, whether about which people to collaborate with at work or what book to read, movie to watch, or hobby to try, our brains evaluate whether the options before us feel like "me" or "not me."

Making choices that feel congruent with our sense of who we are feels rewarding, and decisions that don't fit our sense of self can be more challenging. The value calculation tends to favor choices that the brain interprets as feeling like "me." This makes self-relevance a powerful force in shaping how we and others make decisions.

So how do our brains determine what is "me," and why does it feel so rewarding to make choices that align with our identities?

## ME, NOT ME?

When asked to describe what she is like and how that is reflected in the character she created in Marcel, Jenny describes what she calls a kind of "utilitarian positivity." Utilitarian because they are both resourceful, hard workers. Positivity because they are both optimists with a positive outlook, capable of seeing many possibilities (recall, Marcel can't have a dog, but he can imagine one from a piece of lint!). "That's like my dogma," she says. "And in the Marcel movie, that's how he's living." They also both really want to connect with others. Watching Marcel, we see a piece of Jenny.

In addition to being impressed by Jenny's creative capacity to convey who she is through her art, as a neuroscientist I'm impressed that her brain, and ours, can answer questions like "who am I?" in the first place. Consider all the different operations that Jenny's brain needs to perform to answer this seemingly basic question about what she is like and how it relates to Marcel. To figure out what she is like, Jenny might call to mind memories from her life, memories of creating Marcel, and stitch these together to answer the question. What's going on in her brain when she does this?

In one early study investigating how the brain tracks self-relevance, volunteers reflected on their traits, their current thoughts and feelings, and their physical attributes. During some parts of the brain scan, they thought about their personality traits (Am I . . . Intelligent? Messy? Neurotic?). This type of "trait judgment task" is used frequently to identify brain regions that track self-relevance. During other parts of the brain scan, the volunteers rated how they were feeling in the moment (Am I . . . Bored? Interested? Happy?), and finally, considered their physical traits (Am I . . . Tall? Freckled? Muscular?).* All these forms of self-reflection activated an overlapping area of the medial prefrontal cortex, suggesting that there is at least some shared neural infrastructure that supports our ability to consider who we are in a moment and who we are more generally.

In addition to helping us answer basic questions like "Am I polite?" and "Am I messy?," parts of our self-relevance system, like the medial prefrontal cortex, also track how *important* these different parts of ourselves are to us. Some traits are more "core" to who we are, and other traits depend on the core. For Jenny, her desire to connect with others and her trust in them are core. Her other traits (like sweetness) follow.

When people think about themselves and what they are like, activation increases in the brain's self-relevance system, including the medial prefrontal cortex and regions known as the posterior cingulate and precuneus.

These same brain regions also help us record and call up memories of our experiences (one of the key ways we learn about ourselves) and allow us to imagine the future (which helps us pursue self-relevant goals). All of these kinds of thoughts are important for deciding whether things are relevant to us or not.

---

* To figure out which parts of the brain belong to which system, neuroscientists compare how the brain responds to different tasks that are similar in a lot of ways but differ in how much they might require the particular system the neuroscientist is trying to understand. For example, by comparing what happens when someone decides if "intelligent" describes them versus what happens when they decide if it is written in lowercase, scientists can tell what brain regions track self-relevance, while controlling for what brain regions allow a person to see, read, and understand the words.

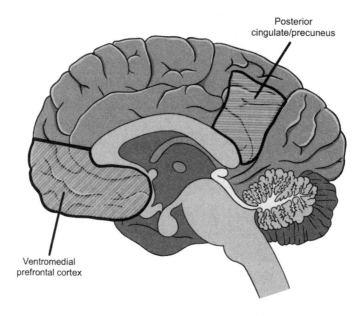

Posterior
cingulate/precuneus

Ventromedial
prefrontal cortex

*The ventromedial prefrontal cortex, posterior cingulate, and precu-
neus, shown here, are involved (along with other regions) in calculating self-relevance, or what's "me" and what's "not me."*

In fact, the self-relevance system filters our experiences according
to their personal significance and emotional intensity. In a research
study where people were asked to remember different things that had
happened in their lives, activation within the medial prefrontal cortex
kept track of the personal significance of the memory, not just what
happened in some objective terms.

Digging a layer deeper, this type of brain imaging research also
highlights how different parts of this system keep track of different
aspects of self-relevance. For example, remembering things that have
happened to us in the past (remembering autobiographical memories
in terms of where we were, who was there, what happened, and so
on) and making meaning of what has happened to us (thinking about
what the event says about our personality, how we've changed, or how
this relates to other significant life experiences we've had) rely on different parts of the brain's self-relevance system.

Different parts of the self-relevance system are also activated to keep track of how good or bad different events are for us and how vividly we think about them. One study showed that imagining things that might happen to us in the future engages the brain's self-relevance system.* When people imagined good things (like winning the lottery) and bad things (like their house burning down), portions of the self-relevance system, including the medial prefrontal cortex and posterior cingulate, tracked how positive or negative the events would be for them. Other parts of the self-relevance system, like the precuneus, increased their activation when the volunteers imagined scenarios more versus less vividly. I find it helpful to know that different parts of my brain handle the valence and vividness of thoughts because it reminds me that I can dial these processes up or down separately.

It is worth pausing for a moment to appreciate what a tool like fMRI brain scans adds to the science here. Without fMRI brain scans, it would be hard to tell which of these processes are related to one another, and in what ways. Brain scans can allow scientists to figure out what kinds of thoughts and feelings that might look different on the surface are actually using similar underlying brain processes, and what things that look similar on the surface are actually different inside our brains. In other words, it isn't obvious that the process of calling memories to mind (remembering the concrete details of where you were, who was there, what happened, and so on) and making meaning of them (thinking about how this relates to other life experiences and what it says about you as a person) draw on different parts of the self-relevance system. On the other hand, thinking about your traits is supported by similar brain regions to thinking about your autobiographical memories and also thinking about the future, or making choices overall.

---

* As we'll explore later, we naturally experience the present as the most vivid, and we need to work to imagine the future vividly, which tends to make it harder to prioritize future goals.

## YOU ARE WHAT YOU LIKE

Another remarkable thing about self-relevance that brain imaging makes apparent: it overlaps heavily with the value system. Calculations of self-relevance and value both rely heavily on activation within brain regions such as the medial prefrontal cortex. In fact, self-relevance and value are so deeply intertwined that they can be hard to disentangle, even in an experimental setting.

I once ran a series of studies with my former graduate student Christin Scholz, now a professor at the University of Amsterdam, and neuroscientist Nicole Cooper, trying to identify regions in the brain that are involved in tracking only one or the other. If we could distinguish between them, we thought it might help us figure out when and why these different processes lead to behavior change and what kinds of interventions tap into each component process. Using the same kind of trait judgment task described earlier in this chapter, we

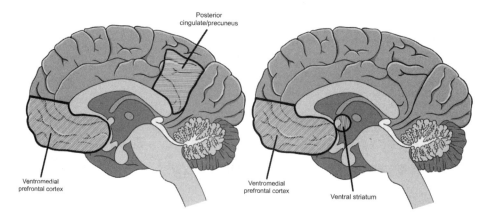

*Key components of the self-relevance (left) and value systems (right) overlap in the medial prefrontal cortex and other regions, such that patterns of brain activity evoked when people make judgments about what they like can also be used to predict self-relevance and vice versa.*

planned to scan people's brains while they decided whether a series of words (like intelligent, polite, lazy, messy, and so on) described them or not—in other words, while they thought about self-relevance. We would also scan them while they thought about whether the same set of words described something good or bad—that is, while they made judgments about value. By comparing the scans afterward, we would be able to see where brain activity was distinct to self-relevance, and where it was distinct to value.

The problem was that it was very difficult to find words for which the volunteers' judgments about self-relevance and value weren't in lockstep. They consistently rated things that described them as not just self-relevant but more valuable. For the experiment to work, we needed to find words for which people would say that something was good, but "not me," and vice versa. It took six rounds of preliminary experiments before we finally came up with a list of words where enough were rated as each possible combination of "me" or "not me" and "good" or "bad" (me + good, me + bad, not me + good, not me + bad). We filled in the "good"/ "not me" quadrant with traits like "aristocratic," "limber," and "wealthy." People also conceded that they were "cynical," "shy," and "impatient," even though these traits aren't typically rated as "good." But even then, the brain patterns that tracked people's assessment of the traits' self-relevance were still really similar to the brain patterns that tracked value.

This tendency of our brains to conflate self-relevance and value is so consistent that another team of researchers found that the same brain patterns evoked when people looked at positive and negative images (think of a cute puppy versus a bloody car crash) were also evoked when people thought about themselves versus others (for example, "Am I polite?" versus "Is my best friend polite?").

But you don't need access to a brain scanner to observe the consequences of this effect. Think about what you would say if you were asked to describe yourself. If you're like most people in Western cultures, your answer will likely include a list of things you like and dislike. *I am a person who likes to go skydiving; I delight in a fistful of arugula as a snack in my office; I love watching comedies where people are kind to each other; I don't like loud parties.* You can find people on dating apps who describe themselves as "loving long walks on the beach and Beethoven" or who "care deeply about climate change" or enjoy "texting you nonsense at all hours of the day and night and [are] weirdly attracted to big eyebrows." We so closely associate preferences and identity, what we value and who we are, that based only on these few bits of information, you probably feel you have some sense of what those people are like. Indeed, in dating profiles and social media accounts, we share our likes and dislikes as a performance of who we are.

Your description of yourself will probably also include some of your core traits. When asked to describe what she is like, for instance, Jenny mentions positivity, resourcefulness, and a desire to connect with others. You might say that, at your core, you are kind, curious, or fair. Interestingly, most people don't choose negative words to describe their core traits.

Why is that? For better or worse, research suggests that most of us have positive illusions about ourselves. We tend to think of ourselves as "above average" (for example, better drivers than average, smarter than average, and so on), even when we are not. In a classic study illustrating this tendency, students rated how well different personality traits described them, as well as how well those traits described the average college student. On average, the students rated themselves higher on desirable traits (for example, cooperative, considerate, respectful) and lower on undesirable traits (for example, deceptive, snobbish, spiteful) than the average college student. This was particularly true for traits that the students believed a person could control (are you cold? deceitful? friendly? loyal? sincere?) and less true for traits that people believed are less under personal control (for example, are you creative? mature? forgetful? bashful?). Most of us don't bias the way we think about ourselves on purpose (exceptions might include job interviews, meeting a new significant other's family, or creating a dating profile), but when the self-relevance system produces an answer to the question "Are you kind?" the answer it gives is optimistic, biased, and only partially complete. And this matters.* Remember the overlap between the self-relevance system and the value system? We use this flawed, oversimplified rendering of ourselves to make decisions.

## "ACADEMICS MAKE GREAT RUNNERS"

As it turns out, identity is a major input to the value calculation. We often say that our choices and actions define who we are, but it goes the other way too: our self-conception also drives our choices.

Researchers have demonstrated this effect in studies exploring how personalized messages, tailored to a person's identity, goals, or values, are more effective at getting someone to take action, compared with generic ones. This type of "message tailoring" can include personalizing messages to suggest that they are designed for a specific person or audience—for instance, adding explicit cues like the person's name ("Jessica, special offer just for you!") or using a messenger who comes from a similar demographic group as the target (my own social media feeds are full of ads with images of white women my age, wearing clothes I might wear). This type of personalized messaging is typically more effective than generic messaging, but messages that more deeply reflect a person's specific needs, values, and goals are even more effective. For example, a randomized trial among thousands of smokers found that smokers stayed away from cigarettes more success-

---

* Imagine a world where all the entrepreneurs, athletes, restaurateurs, and others in lines of work with extremely high failure rates had realistic self-views. Many breakthrough innovations and businesses we enjoy might never have been created, because their inventors and founders might never have tried. (It is also true that we might have fewer failed businesses if people were more able to calibrate a realistic view of what they can do.)

fully when they got messages that took into account their personal smoking history, motives for quitting, and challenges they expected to encounter while quitting, compared with messages that offered more generic advice. After six weeks, 29 percent of the people who received personalized messages had been able to avoid all smoking, whereas of the control group—the group in the study that got generic messages— only 24 percent did. After twelve weeks, 23 percent of the folks getting personalized messages were still successful, compared with 18 percent in the control group.

But how does this work in the brain? A study led by Hannah Chua and Vic Strecher, at the University of Michigan, showed that messages personalized to the smokers' own motivations for kicking the habit increased activation of the self- (as well as social-) relevance system, compared with generic messaging. The researchers recruited ninety-one smokers who were interested in quitting within the next thirty days and developed tailored and untailored coaching messages that the volunteers would see when their brains were being scanned. Based on earlier survey responses, a personalized message for someone who was cost-conscious might remind that volunteer, "You want to quit because you are tired of spending your money on cigarettes." More generic advice for this person might say, "Smokers are admitted to the hospital more often than nonsmokers." On the other hand, a smoker who had expressed stronger concerns about their health might get personalized messages like "You feel your health somewhat limits you, including even taking the stairs," while the generic version for this person might say, "Many smokers quit because they are tired of spending money on cigarettes." In other words, across people, similar messages might count as personalized or generic, depending on their specific concerns and motivations.

Brain imaging showed that the personalized messages activated key parts of the self-relevance system—specifically, the medial prefrontal cortex and precuneus, which are involved in tracking whether something is "me" or "not me"—more than the generic messages (although generic messages also activated these regions). Crucially, the researchers found that people whose self-relevance systems were especially active in response to personalized coaching were more likely to quit over the following four months.

The choice a person is facing doesn't even have to be as big as quitting smoking. For those who think of themselves as a person who follows their heart, attention to how a backpack might make them feel ("offers endless possibilities to express your personality and to be surprised by unique and innovative solutions . . . has become a symbol of discovery, euphoria, and freedom") increases activation in the self-relevance and value systems more than focusing on more objective qualities ("very handy and comfortable thanks to the many internal pockets that allow you to carry everything you need"). The reverse is true for those who think of themselves as being more focused on facts and rational decision-making.

The power of self-relevance means that intentionally focusing our attention on the ways that a choice is consistent or inconsistent with our identity can influence the value calculation and motivate changes to our day-to-day behavior. The most obvious examples of this are often rooted in advertising, politics, or public health messaging, but we can all use this knowledge to be more persuasive by focusing on things we know are important to someone's identity or to notice when others are doing this to us. For example, if someone you mentor at work doesn't think of themself as particularly book smart, but does think of themself as the life of the party, you can highlight how making a big presentation at work is similar to engaging a crowd. Likewise, you can use this as a tool to persuade yourself to do things you want to do but have a hard time actually doing. Say you're an extrovert who has been having a hard time getting outside and moving your body—even though you think you'd feel better if you did. Focusing on how group hiking may be compatible with your self-conception as a sociable person (or as an activity for some alone time if you're more of an introvert) may make you more motivated to hike, especially if you don't see yourself as a physically active person.

Yet it's important to recognize that what is "me" and what is "not me" isn't absolute. Our selves are constellations of traits and identities. I think of myself as a hard worker, a loyal friend, a careful scientist, a patient parent, and a supportive boss. Sometimes I'm also a goofball, an absent-minded professor, an impatient parent, and an oblivious boss. Harvard psychologist Ellen Langer points out that depending on how we look at things, the exact same choice can seem totally compatible or incompatible with who we are. A person might happily see herself as "reliable, spontaneous, and trusting," while a less generous observer might judge her to be "rigid, impulsive, and gullible." Am I the kind of person who might dress up as a hammer when giving a scientific talk on Halloween? Yes . . . it turns out I am.

We can find power in this flexibility. By identifying and highlighting ways that different choices we want to make can be consistent with who we are, we can feel a greater sense of agency and autonomy over our behavior and experiences—and help others do the same. For example, I have never thought of myself as an athlete. When I was a kid, I was good at school. I loved math and science but was not particularly good at sports. I started jogging as an adult as a way to blow off steam and de-stress, but I didn't think of myself as a "runner" and never had ambitions to be fast. Then, one day, my brother Eric (who is a natural athlete and a runner) came to me with a pitch: with some targeted workouts over a month or two, I could get faster, and eventually, running faster would come to feel as easy for me as running at my slower pace.

I was surprised. Why would he think I'd want to do that? Athletic performance had never been important to me, and as my brother, he knew that. But, as my brother, he also knew what *was* important to me. "You know, there are lots of academics who are great runners," he began. "Being a good runner and being a successful academic both require putting in a lot of effort to reach your goals. You've got all the mental skills already."

I smiled. Naturally, I noticed what he was doing: he was encouraging me to see what he thought was consistent with a core aspect of my identity ("academics can be great athletes because they have focus"), rather than opposed to it ("I'm a nerd, not a jock"). But even though I knew what he was doing, it worked. The next morning, I joined Eric and our sister (also a more serious runner) at the track, and over the next week or so, I pushed myself.

When my brother called attention to the ways that training to be a faster runner is compatible with my identity as a hard worker, he made me feel more capable of achieving that goal and more motivated to try. Even though I haven't kept up a highly intensive training regimen without my siblings in town, I do often add a sprint at the end of my run now, knowing what I'm capable of, and inching toward a slightly faster version of myself. So there's a feedforward cycle—the more I make these small moves toward being a better runner, the more it feels like part of "me." This is consistent with what the psychologist and neuroscientist Elliot Berkman refers to as the identity-value model of self-control. When we find a way to make the things we want to do feel compatible with who we are, they fall into place more naturally, and when we can align our sense of self and core values with what we do on a day-to-day basis, we have a sense of agency and autonomy. Think of the way Jenny felt making Marcel.

Self-relevance can be a powerful tool in persuading people (including ourselves) to do things we might not be otherwise inclined to do. This shows up in media that persuade us, feelings of connection with friends who "get us," and in self-talk as we pursue goals. We also need to be aware of this as we scroll through messages intended to tap into our sense of who we are and that are also slowly changing that identity in the process. Particularly as we spend more time online, where the financial incentives of tech platforms are often not aligned with our well-being—where there are large incentives to spread disinformation and where AI can generate large quantities of tailored messaging— we'd do well to stop and notice the extent to which messages are taking advantage of and tapping into our bias that things that seem "like me" are also correct and good. As useful as it is for our self-relevance and value systems to boost our self-esteem, there are times when they can hold us back—especially when changes we're contemplating seem to threaten core elements of our identity.

## DECOMPOSING IN SERVICE OF
## A REALLY GOOD TRANSFORMATION

Jenny Slate experienced one transformation when she realized she could do comedy on her own terms, altering her career to pursue projects that resonated with her sense of self and gaining a deep sense of fulfillment in the process. But later, she faced another, different kind of transformation. She fell in love with a writer and artist named Ben, and in 2020, they learned they were expecting a daughter.

This is when I happened to meet Jenny. Soon after the COVID-19 pandemic took hold, Brett and I temporarily relocated to Massachusetts to spend some time at his mom's house while our kids' school was closed and Penn, where we are both faculty, went virtual. One day that spring, we ran into Brett's childhood friend Ben in a sandy parking lot near the salt marsh where we had come to walk. When Ben introduced us to his fiancée, Jenny, my kids—instead of politely saying hello or "nice to meet you"—sprinted behind our car, kicking up dust and gleefully cackling in their hiding place.

Jenny laughed, too, and joked about their excellent social distancing skills. I didn't know that Jenny was a famous actor then or even the next few times we saw them (though I did think she looked strikingly similar to Mona-Lisa Saperstein, her character on *Parks and Recreation*). As I got to know her in the months that followed, I came to admire how clear she was about her sense of self and what she valued. As comes across in her work, she leads with her heart and is open with her love and delight.

She also doesn't shy away from talking about hard things. In the late fall of that year, as her due date drew closer and she and Ben prepared to leave Massachusetts for Los Angeles, she sent me an email. "I'm starting to feel some sadness and anxiety edge in as we really only have one more week here," she wrote. "There's a central fear: What if I'm a worse version of myself after the baby comes? Or: what if I don't exist anymore?! I'm sure that won't be the case, but sometimes the fear pops up and I guess I just have to look right at it."

I told Jenny that I had felt like that, too, in moments of big change or transformation—moving to a new place, becoming a parent, losing a parent. Maybe you've experienced these feelings as well. When your identity is closely tied to specific domains of your life—your job, romantic relationships, friends, hometown—losing them feels like more than just a change in part of your life; it may feel like losing yourself. Maybe you're contemplating ending a long-term relationship that's no longer working, but you stay in it longer because you can't imagine what "single you" looks like. Or maybe you're considering a career pivot out of a job you don't enjoy. Many of us invest a lot of our identity in our jobs. Who would you be if you weren't a top sales rep anymore?

At the end of her email, Jenny added, "One thing I've got going for me, as I look back, is that at some sort of core level I am consistently myself." In her anxiety, she took comfort in the feeling that there was a core self that was stable, that was *her*. This is true of most of us. Having a relatively fixed sense of self helps us organize our experiences and see where we fit within the web of relationships and experiences we've had.

But holding onto the idea of a fixed, core identity can make changes seem scarier as well. Emphasizing a stable but perhaps rigid or outdated sense of self may give us the impression that this will never, and *can* never, change. Jenny's strong sense of identity gave her comfort, but worrying about how that might alter also became a source of anxiety ahead of a major life transition. If we think of "me" (as I have been) as "good," then what happens when we consider change?

Recall that the version of our "self" that the self-relevance system conjures up is more like a dating profile than an objective and true lens on who we could be. When we're facing a change, one approach to smooth the transition might be to think about ways the impending change is consistent with our sense of self—like when my brother Eric helped me think of myself as a hardworking runner. It can also be helpful to try to look beyond those constraints—to let go of prior ideas about who we are, or at least not hold so tightly to them. For example, Eric drew my attention to the parallels between being an academic and being a runner and helped me challenge my own idea that one thing precluded the other. We can also do this for ourselves; even though I didn't think of myself as an athlete as a kid, I might ask myself whether the things that made me feel that way then still apply or whether I even think a person needs to identify as an "athlete" to move their body and get the benefits.

In other words, there are situations when it feels great to do work that feels authentic to ourselves (like Jenny's experience in creating Marcel). But there are also situations in which it can hold us back from change by constraining who we think we are or what people "like me" are supposed to do. In addition, structural barriers or discrimination may prevent us from aligning our behaviors with who our authentic selves are: if we break away from traditional conceptions of how we should behave, we may face backlash. The latter is often in service of social change, but it's not without consequences. Still, we can and do change, and what we think of as "me" can be more capacious than we might initially give ourselves credit for. Just as we can expand the way we think about what might be rewarding, we can also work to imagine more qualities available in our repertoire.

Meditation is one tool that can help us let go of rigid ideas of who we are and what we are capable of. People who meditate a lot, like Buddhist monks, have brain patterns that look different from most of ours, particularly when it comes to brain regions that construct our sense of self. Some research suggests that exercises like meditation can make us more open to new ideas about who we are or could be. For example, transformative experiences that come from years of meditation practice often include the sensation of a person's core sense of self no longer being separate and unique. In fact, research suggests that long-time meditation practitioners use their medial prefrontal cortex differently. During resting states (when they aren't exposed to any stimulus in particular), meditators show greater connectivity in this part of the brain than nonmeditators. Even short interventions that teach people to focus on the variability of their traits and how they are feeling in the moment can quiet the medial prefrontal cortex and shift how it communicates with other brain regions. Psychedelics have similarly been shown to reduce activation of the medial prefrontal cortex and have been associated with "ego dissolution," which refers to a letting go of the restricted sense of self.

In line with these ideas, research led by Princeton neuroscientist Molly Crockett highlights how transformative experiences are characterized by this kind of expansion. Based on research at festivals like Burning Man, Molly's team concluded that many transformative experiences involve greater and broader feelings of connectedness to other people, which are typically constrained by the boundaries of "me" and "not me" that our brains generate. In other words, taking a psychedelic drug, or experiencing community in a radically different way, can help us let go of that strong sense of ego that overlaps with so many of the decisions we make. We can also practice making those boundaries more permeable through day-to-day activities like meditation and other contemplative practices, and we can also notice and harness the ways we are connected to other people.

Many experiences that enhance or threaten our relationships with others can be transformative. Becoming a parent, losing a loved one, falling in love, connecting with a teacher or friend who helps us see the world in new ways—each of these can help us organize, and in some cases rearrange, what we know about ourselves, how we make decisions, what we value. This, in turn, shapes the choices we make and who we become.

Personally, I try to consider that my brain's self-relevance system can adapt, and will adapt; that I don't have to believe everything that first comes to mind about who I am; and that I can understand the ways that social norms and societal expectations influence both my sense of self and value calculations. Paying attention to this feels like a license to imagine other possibilities and transformations as well. Could I be the kind of person who is a skydiving instructor? A mathematician? A venture capitalist? Examples in the media and broader societal norms have shaped my brain's immediate stereotypes about who does these jobs, and these societal constraints are real (in shaping opportunities and what it might be like if I entered those professions). But that's different from whether I can imagine myself being the kind of person capable of doing these jobs.

On an August afternoon almost two years after Jenny's daughter, Ida, was born, we sat cross-legged in her hilltop writing cabin. With the windows and doors wide open to the bright blue sky, and a breeze dancing with the leaves of the trees surrounding us, I asked her about the day she wrote me that email, the one in which she wondered if she would still be herself after becoming a mom.

She lit up, cocked her head thoughtfully to the side, and remembered what she called a "dreadful anticipatory energy," but also identified it as the feeling people get right before a big change, "right before they are about to sort of, like, decompose in service of a really good transformation," and laughed softly. She described the fears she'd had about becoming a parent and what it would mean for her relationship with Ben and for how she would see herself. Recalling times when her parents seemed unsatisfied with parenting, for instance, she remembered worrying that "this thing that I'm about to do is bigger than me and it's going to destroy me."

But after giving birth and seeing what she and the people in her life were able to do, she felt a new and expansive sense of power. "Seeing what I was capable of, giving birth, seeing what my partner was capable of in supporting me through that, seeing the baby herself and the entire catalog of new feelings that come with that, that are so pure and so good . . . I, all of a sudden, was able to realize that, yeah, at the core of me is love, is a big power to make great connections with people that I admire."

These new feelings, experiences, and connections to others have made her feel more like a leader, a person with power to connect with others. And those connections have transformed her into who she is right now.

Jenny's creation of Marcel was one part of her exploration of what it was like to do things that were "like her," and becoming a parent was another; her self-relevance system helped identify what those were and make meaning out of them. But as a highly social species, we also determine who we think we are by what other people see in us. The warm reception that millions of people gave Marcel was rewarding and reinforced Jenny's sense that it was valuable to lean into doing work that was compatible with those parts of herself. And Jenny is not alone in this. Many of us use information about what we think others think about us to figure out who we are. This feedback can push us toward or away from different paths. But how is it, exactly, that other people shape our thoughts, feelings, and choices? That relies on another, closely related, brain system, the social relevance system.

# 3

# Who Are We?

TABITHA CARVAN LOVES Benedict Cumberbatch. She loves him so much that she wrote an entire book about her passion—though, officially, that's not what the book is about. *This Is Not a Book about Benedict Cumberbatch* is about the joy of getting so into something (in Tabitha's case, a British actor best known for playing Sherlock Holmes and the Marvel character Dr. Strange) that it consumes you.

I was given a copy of Tabitha's book by my friend and fellow neuroscientist Rebecca Saxe. When I started it, I had no particular interest in Benedict Cumberbatch. I wouldn't have called him one of the sexiest men alive, as no less an authority than *Glamour* magazine has, and I had never watched *Sherlock*, the television series that made him famous and first spawned Tabitha's obsession, along with countless others'. If I'm being honest, if I thought anything about his looks, it was that they were plain, with his wide-set and upward-tilted eyes sitting on his unusually long face. But after reading over two hundred pages of Tabitha's ode to each inch of him, I went online to have another look. The result: I could maybe see it, but I still wasn't sure.

The next morning, I mentioned the book to a friend while we were on a morning run together. It was late July in Massachusetts, the sun having just breached the shaggy tops of the dense trees that give shade to runners who aren't ready for the day's heat. I was feeling the delight of being in sync with my friend, who had come running out of her parents' house just as I approached the driveway and kept running with me at exactly the right pace to be able to talk. And then, at my mention of Cumberbatch, she stopped in her tracks.

"Oh yeah!" she said, her face lighting up. As she started gushing, I realized she shared Tabitha and *Glamour*'s view: she was super into him. Pausing here to witness her ebullience, I became even more curious: what could I be missing?

That night, after the kids went to bed, I told Brett that I wanted to watch *Sherlock*. Unbeknownst to me, he had already seen the entire series, but said he was happy to watch it again. As he prepared us a salad at the counter, I asked him to elaborate. (How had I missed this Cumberbinge? Was Brett, too, enamored?) He replied, "Well, Sherlock is a hot *character*."

And as we watched the first episode together, I'll admit I began to find Sherlock's competence and attention to detail peculiarly charming. But what really got me was watching Cumberbatch interviews online. In one, he is asked how he feels about being named "hottest person this summer," to which he replies, "It just makes me giggle." In another, he jokes self-deprecatingly with a BBC host about the time he voiced a nature documentary and repeatedly mispronounced "penguins" (as "pengwings," "penglings," and other variants).

In this instance, other people's opinions challenged me to look further and opened me up to learning about his charm, his humor, and his pengwings. In the span of a few weeks, I had gone from not thinking much at all of Benedict Cumberbatch to agreeing with (apparently) many others: he's attractive. But why should the opinion of an author in Australia, or the people she interviewed for her book, or the readers of *Glamour*—all faraway people with whom I'll likely never interact—change how I feel about Benedict Cumberbatch—another faraway person with whom I'll likely never interact?

Benedict Cumberbatch himself may lie outside my core expertise, but understanding why others' enthusiasm affects mine *is* squarely in my wheelhouse. Neuroscientists like me study the brain mechanisms by which what other people think—or, more precisely, what we think other people think—changes our own opinions and actions. Put another way, we study how the brain's assessments of the *social relevance* of a choice can powerfully shape the value calculation.

In the years leading up to the release of *Sherlock*, in fact, neuroscientists were investigating exactly the question that I would later ponder—that is, how social relevance can shift the way people evaluate other people's beauty. In a study published in 2009, researchers in the Netherlands, led by Vasily Klucharev and Ale Smidts, scanned twenty-four young women's brains while the volunteers evaluated pictures of over two hundred other women's faces. The participants first rated how attractive each face was, then were told how peers at universities in Paris and Milan had ostensibly rated the same faces. Then, after emerging from the scanner, the volunteers rated the same faces again.

The researchers wanted to see whether the volunteers would change their ratings in response to their peers' views, and what happened in their brains when they learned about their peers' opinions. Importantly, the "peer" ratings weren't actually generated by peers; they were generated at random by a computer. This gave all the faces an equal chance of being labeled more or less attractive than participants had initially rated them. In this way, the scientists could ensure that any shifts in ratings were due to peer influence, rather than the underlying, "objective" attractiveness of each face.*

The study revealed that the brain's value system not only detects

---

* In some of the studies we'll encounter in this book, researchers use deception, such as providing false feedback to participants, in order to maintain the validity of the study results. For example, here, deception is used to make sure that the effects of social influence are not confounded with other aspects of the stimuli, such as facial features that people might already think of as more or less attractive. See the endnote for more on why, and how researchers seek to minimize harms of deception.

when our opinions are misaligned with others but also helps bring us back in sync with the group. When the volunteers learned that their views about a given face diverged from their peers', activation in their brain's value system initially plummeted, regardless of whether they had found the face more or less attractive than others supposedly had. Like an alarm, the value system registered when the volunteers were out of alignment with others—much like the negative prediction errors we saw in Chapter 1, when the brain made note of when a choice was less rewarding than anticipated. In fact, the more strongly a participant's value system showed this prediction error response, the more likely they were to change their ratings of the faces to align with the group later. Peer feedback triggered the same kind of prediction errors that help us learn from our own experience.

But had the participants in the study really changed their minds? After all, as I learned that more people loved Cumberbatch, I had initially nodded away as they gushed. It's common behavior to publicly conform to peers' viewpoints while privately not changing our underlying opinions. A couple of years after the initial study, a team of researchers at Harvard ran much the same experiment, this time scanning participants' brains while they made their second facial-attractiveness rating so the scientists could see what was really happening in the participants' value systems after they were exposed to the peer feedback.

It turned out that after the volunteers learned which faces peers thought were more and less attractive, both the volunteers' survey rating responses and activity in their value systems tracked with the peers' ratings. They weren't just publicly conforming; their underlying value calculations were updated by their peers' judgments—which, it's worth repeating, were assigned at random and had nothing to do with how attractive the faces really were. In the same way, when I watched *Sherlock*, my friend wasn't in the room, neither was Tabitha, and nothing about Cumberbatch's face had changed. But as I watched him squint his eyes as he put the pieces of a complicated mystery together, I began to think to myself, "Yeah, okay, I see it." This study illuminated why: our ideas about what our peers think—or social relevance—act on the value system to influence what we see as beautiful or valuable.

Since these early studies, other neuroscientists, including folks on my team, have found that peers' opinions influence our value calculations about much more than other people's attractiveness. From what foods we want to eat, to what products we would recommend, to something as seemingly personal as what art we display on our walls, what we think other people think guides our judgments.

Social psychologists have known for a long time that other people's opinions can change what we do, and we often think about social influence as something that happens "out there," acting on people from the outside. As a corollary, we judge shifts toward conformity as being inauthentic, untrue to ourselves, or otherwise artificial. But psychology and neuroscience research suggests that this is a mistake. When we pay attention to what other people think and do, it can change not only our outward behavior but also our own value calculations, privately held beliefs, and who we are.

Our brain's social relevance system helps us figure out what others think, feel, and do, and this can change both what we value and find enticing and what we actually *do*—including not just my opinions about Benedict Cumberbatch and whether to watch *Sherlock* but also whether to vote, pay taxes, exercise, and more. Drawing attention to what other people think and feel can be a powerful force to purposefully change our own decision-making and that of others, as well as to guard against its potential negative effects. But for any of this to work, we need to know what others are thinking in the first place.

## YOU READ MY MIND

When my kids, Emmett and Theo, were about five years old, they loved playing a game we call "You Read My Mind." Here's how to play: stare into another person's eyes and ask, "What am I thinking about?" Your partner then guesses, for instance, "Are you thinking about unicorns?" If you're not, you can offer a hint: "No, not unicorns, but something related to unicorns." They can then try again— for example, "Fairies?!" If they are right, you exclaim, "You read my mind!" and everyone delights.

You might never have played this exact game with a small child, but most of us play a version of it every time we interact with another person. Consciously and unconsciously, our brains make predictions about what the other person is thinking and feeling, based on our past experiences with people generally, things we know about this specific person, their facial expressions, how they're acting, the situation they are in, and myriad other factors. This process—which scientists call "mentalizing" or "theory of mind"—allows us to understand and simulate what others might think and feel, which in turn lets us guess what might happen next and how it might affect us.

My friend Rebecca (yes, the very same one who gave me Tabitha's book) is one of the world's leading experts on the neuroscience of mentalizing and theory of mind. In a series of studies run since the early 2000s, she and her team identified specific regions in the brain involved in these processes. Although scientists technically distinguish between mentalizing and theory of mind, I'll refer to these regions collectively as the "social relevance system." In the pictures shown here, you can see some of the key regions that help people understand what others think and feel. You might notice that some of these regions overlap with the regions we saw are involved in valuation and self-relevance processing (see Chapters 1 and 2).

How do we know what brain regions support thinking about thoughts and feelings? In one early experiment, Rebecca and an MIT colleague, Nancy Kanwisher, looked at what happened in people's brains when they made inferences about what others might think and feel, compared with other types of inferences. Volunteers in their study read miniature stories that first prompted them to think about other

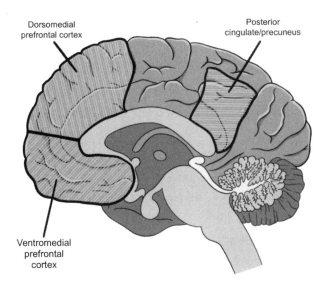

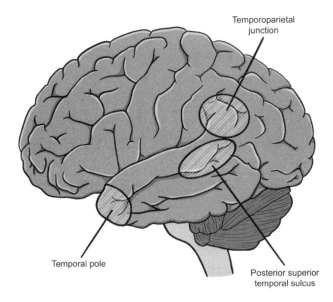

*The posterior cingulate cortex and the precuneus, the ventromedial prefrontal cortex, and the dorsomedial prefrontal cortex, along the middle section of the brain, and the temporoparietal junction, posterior superior temporal sulcus, and the temporal pole, on the sides of the brain, are all regions implicated in "mentalizing," or thinking about what oneself or others think and feel.*

people's thoughts and feelings. For example, what thoughts come to your mind when you read this story?

"*A boy is making a papier-mâché project for his art class. He spends hours ripping newspaper into even strips. Then he goes out to buy flour. His mother comes home and throws all the newspaper strips away.*"

The story itself doesn't say whether the kid cared that his mother threw the materials for his project away, yet you probably have a guess. Did your brain spontaneously infer that the boy might feel upset? Chances are you also have a guess about what he'll do next: maybe he might feel angry at his mom and ask her to recover the trash, or maybe he'll throw a tantrum.

Now consider this story:

"*A pot of water was left on low heat yesterday in case anybody wanted tea. The pot stayed on the heat all night. Nobody did drink tea, but this morning, the water was gone.*"

This story is similar to the one about the boy, in that it prompts you to make an automatic inference. For example, perhaps you spontaneously inferred that the heat made the water evaporate, just as you spontaneously inferred how the mom's actions might make the boy feel (losing something you worked hard on feels upsetting). But Rebecca and Nancy found that these stories elicited different responses in the brain.

Thinking about other people's thoughts, feelings, and beliefs, as in the first story, uniquely activated a region of the brain that sits just behind and above your ears known as the temporoparietal junction, along with other brain regions that are part of the social relevance system. By contrast, when people thought about things like why there was no water in the pot, or even about other people in a more general way—for instance, about their appearance—these regions stayed comparatively quiet.*

---

* Note that these areas have been shown to activate less in this type of task in people on the autism spectrum, where differences in the activation of these areas are associated with social difficulties. This lends support to the region's role in social mental

Rebecca and Nancy's results suggested that this set of areas in the brain specializes in understanding what others are thinking and feeling. Further studies, by their teams and others, have since solidified the role these regions play in helping us understand what other people think, which is one key ingredient in determining social relevance.

In addition to making sense of what another person is currently thinking and feeling, the social relevance system also helps us make predictions about what other people will think, feel, and do in the future. We do this through a type of *predictive coding* that tracks what we expect will happen and takes note when something unexpected happens. This allows the social relevance system to constantly refine its predictions. Psychologists have also learned that, overall, people's predictions are pretty accurate when it comes to what thoughts, feelings, or behaviors might follow another. This ability is useful if you want to coordinate, cooperate, or negotiate with another person, understand the plot of a story, or even anticipate a rival's moves in a game of chess.

Interestingly, we're not born with the ability to harness the power of this system; rather, it develops over time. For example, Rebecca and her student Hilary Richardson have found that as kids move through childhood, their social relevance systems make increasingly active predictions about others' thoughts and feelings. On average, in adults, you can see the results of fully developed predictive coding when they hear a story or see a movie that they are familiar with. The second time an adult hears the story, they have a sense of what is coming, and so activation in the social relevance system (which keeps track of characters' thoughts, feelings, and motives) comes earlier than when they heard the story the first time.

To find out if the social relevance systems of kids of different ages also anticipate what will happen next in the same way, Hilary and

---

processes and also highlights that brain functions are not universal. People use their brains differently.

Rebecca had both very young children (including three- and four-year-olds) and slightly older kids (six- and seven-year-olds) watch an animated short film, *Partly Cloudy*, twice. The movie follows the adventures of a cloud named Gus, who is in charge of making new baby animals like crocodiles and porcupines, and his delivery stork, Pek, who valiantly attempts to carry the creatures to their parents. The movie showcases a wide range of emotional experiences, from the fear that Pek feels in approaching the prickly animals, to the sadness that Gus feels when he thinks Pek is abandoning him, to the happiness Gus feels when Pek returns.

In Hilary and Rebecca's study, when the six- and seven-year-olds saw *Partly Cloudy* a second time, their brains' social relevance systems anticipated what would happen next, with activation coming earlier than the first time they saw the movie, suggesting that they were anticipating what the characters would think and feel in advance. Activation in the three- and four-year-olds' social relevance regions stayed largely the same the first and second time they viewed the movie—indicating that they had not yet developed the tendency to anticipate what would happen next, socially.

You might be wondering if this simply means that six- and seven-year-olds have better memory than three- and four-year-olds, but the change Hilary and Rebecca observed to the older kids' activity patterns was specific to the social relevance system. There were few significant changes to the patterns of other key brain systems involved in tracking what happens in the movie. This suggests that it wasn't just that the older kids remembered the movie better but that they had a more developed understanding of what others might be thinking and were using this to predict what the characters would think and do next in the movie.

This is one of the reasons it's so much fun to play "You Read My Mind" with a five-year-old. At that age, they are just beginning to be able to more accurately decode others' thoughts and feelings, and it's exciting for them to discover they can figure out what someone else is thinking by asking questions and observing. Frankly, I love the genuine wonder with which kids exclaim, "You read my mind!" And I think it is worth taking a moment, even as adults, to marvel that we can read other people's minds in this way. How did Emmett know I was thinking about fairies?! And why would our brains evolve to be able to make these kinds of guesses so well?

## I WANT YOU TO WANT ME

Crucially, although the social relevance system tracks what other people think and feel about an almost infinite variety of things, our brains seem to be *especially* interested in particular kinds of social information. Beyond stories about kids making papier-mâché and other people's intentions, our brains care a lot about what other people think *about us.*

We want to be liked, respected, and cared for. Evolutionary psychologists theorize that these feelings were essential to survival in humanity's deep past. If you were liked, respected, and cared for, it was more likely that other humans would help you stay warm, defend you against predators, share food, and so on. Conversely, if you didn't care whether people liked you and therefore acted in alienating ways, you were more likely to get cast out or otherwise left on your own, die, and fail to pass on your genes (and your unhelpful tendencies).

We can see this legacy in how we experience a variety of social situations, whether they're objectively high- or low-stakes. At a happy hour with friends once, for example, someone asked me what music I listened to in high school. I froze. I tend to be pretty open, but in that moment memories of judgy high school peers' opinions of my musical taste came to the surface. (I loved the Indigo Girls and Dar Williams, but bands like Dave Matthews and Phish were more broadly popular among the guys who hosted parties on the weekends.) What if revealing my uncool musical preferences made my friends at the happy hour think less of me? I mumbled something about the Indigo Girls, and the conversation moved on.

When I got home that evening, I thought to myself, "This is ridiculous. I'm a fully grown adult. Why wouldn't I want to share the music I liked in high school with my current adult friends?!"

As an adult scientist with the knowledge I have now, I can see more clearly how often what is seen as cool or normal isn't about an objective standard but rather about who has power to shape what is "mainstream." Cultural norms and values are shaped by media industries dominated by men. Maybe this realization fired me up a bit—I decided to make a playlist with some of the artists I loved in high school—Indigo Girls, Mary Chapin Carpenter, Mary J. Blige, Aretha Franklin, Dar Williams, Lisa Loeb, TLC, Cyndi Lauper, Jill Sobule, Jewel, Bonnie Tyler, Joni Mitchell. In what felt, absurdly, like an act of immense bravery, I clicked "send" on an email to the happy hour group.

Like the study participants who were asked to judge attractiveness and then learned that their tastes were out of sync with their peers, in high school I experienced a negative prediction error when I came to think that some of my friends at the time disapproved of my musical choices, and what my value system learned then stuck with me for decades. Indeed, research by UCLA neuroscientist Naomi Eisenberger and others shows that just as experiencing physical pain motivates us to take action to avoid damaging our bodies (think of how you automatically pull your hand back if you accidentally touch the hot burner of a stove), negative social feedback or rejection activates an analogous form of social pain that acts as an alarm system that motivates us to repair our social ties. In other words, this "social punishment," as scientists put it, likely updated how my brain calculated the value of sharing music, which may have made me hesitate at drinks with my friends years later.

Now, as they responded to my playlist, my social relevance and value systems were updating again. "This is so cool!" wrote one friend.

Another made her own playlist featuring some of her favorite high school jams. That list included many songs I also loved, both in high school and now. I felt *amazing*. As these unexpectedly enthusiastic responses came in, it created a positive prediction error that made me more likely to share again in the future (indeed, I'm telling you about it right now!).

The reason I felt so good after hearing from my friends is that, as it turns out, the brain's value system treats positive social feedback like it treats other, more tangible rewards, like chocolate and money. Feeling connected to others also releases chemicals in your brain, including special types of opioids (called μ-opioids, pronounced "mew ope-ee-oids"). You've probably heard of opioids in the context of the drug crisis—people use drugs like heroin, morphine, and oxycodone because they bind to opioid receptors in the brain, dampening pain and activating feelings of pleasure. You might not realize, though, that your brain also manufactures this type of chemical and releases it in a much safer way when you connect with loved ones (or do other pleasurable things, like eat yummy foods). The warm and affectionate feelings we get from being with people we love and care about come from this natural drug that our body produces, and that helps motivate us to maintain the kinds of relationships that are essential for our long-term health. Indeed, research shows that being connected to others releases chemicals that increase activation of the value system and make us feel good.

This effect in the brain also helps explain why social rewards can feel so gratifying. For instance, when he was in first grade, I tried to bribe one of my kids to do his spelling homework by telling him I'd pay him a quarter if he learned the words by the end of the day (he knows that a quarter can buy candy at the general store near our house). He leveled his best negotiating face at me and said, "If I learn all of them, will you make me a certificate with today's date and write 'Good job, Theo!'?" In his mind, a quarter was worth less than proof of my praise and approval.

Social rewards can be a powerful motivator—sometimes on par with money or sustenance. But can we use this knowledge to actually change our, and others', behaviors?

## WORKING TOGETHER

When was the last time you *really* looked at your electricity bill? Not just the "amount owed" part, but the rest of it? Chances are that, like the one I get each month, it shows how much electricity your household consumes compared with your neighbors—and there's a good reason for this. Because the value calculation takes into account social information, just telling me what others are doing can nudge me to change my behavior.

These messages are now extremely common in energy bills across many communities in the United States because research has shown that learning about others' behavior can be an effective way to encourage people to conserve—even though the people in question don't typically realize that the actions of others are influencing their behavior. In one study published in 2008, for example, researchers surveyed over eight hundred Californians about their energy conservation practices, as well as their impressions of how much their neighbors and others in their city and state conserved. When asked what influenced their energy conservation decisions, the respondents generally rated what others were doing lower than reasons like saving money or benefiting society and the environment. But when the researchers then looked at how much energy people used based on electricity meters at their homes, people's perceptions of other people's energy use were the best predictor of their behavior.

Based on these results, the research team next ran an experiment in which they distributed door hangers promoting energy conservation to nearly a thousand households in California. Some of the door hangers highlighted what others were doing ("77% of San Marcos residents often use fans instead of air conditioning to keep cool in the summer"), while others merely encouraged their recipients to save energy in a more abstract way ("How can you conserve energy this summer? By using fans instead of air conditioning!"), or appealed to other motivations that people may have to save energy ("You could save up to $54 per month by using fans instead of air conditioning"). The households that were told about their neighbors' conservation efforts then reduced their consumption by larger amounts than people who received other kinds of messages. However, they still reported what others do as the least important reason for their energy-saving behaviors.

Operating beneath people's conscious awareness, this use of "social proof"—highlighting what others are doing, or, as we might also put it, drawing attention to the social relevance of a behavior—proved so effective that many energy providers now include this type of nudge on every bill they send out, including mine and probably yours. If I see that I am using more electricity than others around me, I may be more likely to try to decrease my own energy use. Paralleling the energy-saving results in California, research has shown that highlighting social relevance can make people more likely to adopt a range of consequential behaviors—from towel reuse in hotels,* to voting, to exercise—even while they remain unaware of how powerfully social influence is affecting their opinions and actions.

These studies collectively demonstrate the power of social relevance. But they don't get at the underlying mechanism. Did the participants in these studies genuinely start to see the world differently? Did they really come to attach greater value to the ideas of conserving resources, voting, and exercising—similar to the way participants in the facial attractiveness studies came to value faces differently?

---

* Note that this was true in an American context, where people underestimate how many other guests reuse their towels. In a German context, where guests already expect that others are reusing their towels, the norm message was not more effective than standard messaging, highlighting the importance of understanding the cultural context of the norm being targeted.

To answer this question, my team, led by Prateekshit "Kanu" Pandey and Yoona Kang, tested how social norms in people's social networks relate to their value calculations. We scanned the brains of over two hundred volunteers in Philadelphia while they were coached about getting more exercise. The volunteers were all people who spent a lot of time sitting, and many were not meeting the federal guidelines for the minimum amount of exercise a person should get in a week (150 minutes of moderate activity such as brisk walking). To find out what kinds of norms the volunteers were typically exposed to in their social networks, we asked each how active or sedentary their friends were. We wanted to know if people who had more friends who were exercisers would spontaneously infer more social relevance from the coaching messages they viewed while in the brain scanner.

In the lab, we monitored each volunteer's brain activity while they viewed coaching messages that suggested different ways people could get more movement and steps in—parking farther away from work, taking stretch breaks, dancing, and so on. Notably, unlike other experiments we've seen in this chapter, these messages weren't focused on conveying normative information; that is, the messages didn't focus on how others in people's social networks were behaving or how cool people seem while exercising. Instead, we focused on how and why exercise might be good for you. Still, we found that the participants' brain responses were indeed related to how often their friends exercised. People who had a lot of active friends also had value systems that responded more positively to the coaching messages, which in turn predicted who would change their behavior and get more exercise in the following weeks, measured with fitness trackers. This highlights one way that social relevance might increase value. The people who saw that their friends exercised may have been primed to see more value in the coaching messages suggesting that they get more exercise themselves.

People with real friends who were active were more receptive to positive behavior change. By the same token, my friends' general adoration of Benedict Cumberbatch probably made me more receptive to Tabitha's not-a-Cumberbatch-manifesto, which in turn ultimately led to me to turn on a TV show, *Sherlock*, I normally wouldn't have watched. In our daily lives, our behavior change is a combination of intentional messages we receive and the influence of people around us. And, although people may not always pay attention to how powerful these influences are, we can practice becoming more aware of the social influences that contribute to our decision-making, deciding what kinds of influences are best aligned with our goals and cultivating agentic alignment. What do you see people around you doing that you respect, admire, or want to emulate? What do you see people around you doing that you want to challenge or avoid?

We can deliberately create and participate in relationships with other people in support of our goals. At biweekly planning meetings for my lab, I like to take advantage of the power of social relevance, social support, and commitment to increase the likelihood that we'll follow through with the things we want to do. Each person lays out their goals for the upcoming two weeks, and the rest of the team gives feedback about whether the goals seem realistic. This has several positive effects. First, seeing how others are prioritizing their well-being can make it easier to explicitly build in time for doing it ourselves. At other times, people bring up a task they don't really want to do but that needs to get done. To make it more rewarding—in other words, to dial up social relevance in their value calculation—other team members will offer to do a "work on that thing you don't want to work on time" session (WOTTYDWTWOT; pronounced "wotty'd wot wot").* They'll get together for a set amount of time, give each other support and encouragement, and hold each other accountable. Having another person there tips the value calculation to make it more enjoyable to get started, and committing to doing it together makes it more likely they'll follow through.

---

* Thanks to Elliot Berkman for this delightful acronym.

But as incredibly well as our brains usually work to parse complex social interactions—and as powerful an influence for good as we've seen social relevance can be—it's also important to recognize when those same systems might be leading us to conform at times when it doesn't serve us, or the world around us.

## THE EMPEROR'S NEW CLOTHES

In the mid-2000s, the Petrified Forest National Park had a problem. The park, an Arizona grassland scattered with the petrified remains of a two-hundred-million-year-old forest, was disappearing—literally, bit by bit. Visitors were picking up the irreplaceable fossils along the trails and taking them home.

The park service knew they had to do something, so they installed signs admonishing visitors: "Many past visitors have removed the petrified wood from the park, destroying the natural state of the Petrified Forest." And when that didn't work, they called in a team of psychologists led by Bob Cialdini, who looked at those signs and suspected they might actually be part of the problem.

The psychologists conducted an experiment where some visitors saw the park service's original signs, some saw signs that simply asked people not to steal, and some visitors saw no signs at all. The researchers' hunch was confirmed: A simple message asking people not to steal reduced theft, compared with no message. But a message highlighting how many other people had already stolen, like the original signs, *increased* theft severalfold. The park would have been better off not putting up any signs at all. Social relevance can backfire.

Just as we've explored the way our friends can serve as positive role models whose actions and opinions help tip the scales of our value calculations, the same can work in reverse. For example, if it rains in Boston and my sister skips her run, knowing this might make me less likely to go for my run and indulge in binging a TV show instead.

This is the flip side of the positive effects of norms on exercise that we explored earlier.

The influence of social relevance judgments on the value calculation can also have broader societal consequences. For example, research led by my former PhD student Keana Richards and Penn psychologist Coren Apicella explored how a stereotype—specifically, the stereotype that women tend to prepare more than men (as when practicing for a presentation or studying for a test)—might become self-perpetuating. Stereotypes are created by the social world we live in and tell us social information: "what I think you think." This information shapes our own perceptions of what is valuable and how we should act, but stereotypes can also push us to act against our best interests.

To understand this better, we recruited volunteers from a website where people can get paid to perform tasks, asked them to do as many math problems as they could within a short time limit (thirty seconds to two minutes, depending on the study), and paid them based on their performance. Before they started, the volunteers could study by doing as many practice problems as they liked, but they were not paid for this time. The men and women turned out to perform about equally well on the task, with or without preparation—but the women tended to prepare more. This was costly for them, since they could have spent that unpaid time doing a different task on the website, for which they would be paid.

Why did the men prepare less and the women prepare more? When surveyed after about how much they thought men and women would prepare for the task, many volunteers said they expected women to prepare more. The extent to which people endorsed this stereotype also correlated with how much they prepared themselves, according to their gender, with women who endorsed the stereotype more tending to prepare more and men who endorsed the stereotype more tending to prepare less. So, ideas about what is socially relevant aren't universal—but when society expects certain things from one group of people more than another, these stereotypes can become self-perpetuating.

The power of social relevance can also cause people to go along with patently false information. Perhaps you know the story of the emperor's new clothes, in which two swindling tailors convince a vain emperor that they know how to weave a fine cloth that is invisible to incompetent people. To avoid looking foolish, everyone agrees, in the emperor's presence and out of it, that the clothes are beautiful. Then the emperor takes part in a parade where all the townspeople are gathered—and a child loudly asks why the emperor has no clothes on.

Although this folktale is hundreds of years old, it captures real-life behavior that we continue to observe, both in the lab and in our daily lives. In the 1950s, the psychologist Solomon Asch conducted a series of famous experiments in which volunteers were shown a line like the one on the left and asked which of the three on the right it matched:

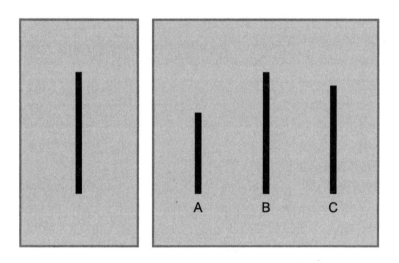

*In the Asch conformity experiments, participants were presented with one line (left) and asked to select which of three other lines (right) was the same length. In critical trials, other participants (who were secretly part of the research team) selected an obviously wrong answer, like A.*

When participants were asked on their own, it was an easy task: the answer was line B—only 1 percent of volunteers gave a wrong answer in this context. But when Asch asked them in the presence of a group of "confederates"—people who were pretending to be volunteers but were actually in on the experiment—who confidently gave an incorrect answer, a third of the volunteers went along with the group. They agreed to an answer that their senses clearly contradicted. And their senses *did* contradict it—their visual perceptions remained unchanged, though they gave answers in contradiction to what they saw.

Many of us would probably like to believe that we wouldn't fall prey to the false line judgments in Asch's study, that we'd proclaim that the emperor is naked, and that we're not influenced by what we see online. We'd like to think we'd stand up and say, "No! Line B is clearly the only line that is the same length as the reference line." But social rewards are very powerful. Recently, when I was in the car with my grandmother and a friend, my grandmother pointed to the sun shining brightly behind some clouds and said, "Wow! Look at that." My friend said, "Yeah, it's crazy how brightly the moon can shine." I was taken aback. It was clearly the sun. When I challenged my friend, she said, "Yeah, I thought that same thing last week, but all of my friends assured me that the moon can reflect light really brightly during the day." I was flabbergasted. I called my friend's boyfriend on the phone and confirmed that her friends had been messing with her—and convincingly so! By insisting as a group that the sun was the moon, they made her doubt a basic fact so strongly that she then passed on the false information to us.

Getting pranked by friends is the least of it. False news spreads farther and faster online than true news, and mainstream media also sometimes amplify misleading information. As my Penn colleague Duncan Watts, who studies social networks and collective behavior, notes, information need not be patently false to leave us with the wrong impression: "Presenting partial or biased data, quoting sources selectively, omitting alternative explanations, improperly equating unequal arguments, conflating correlation with causation, using loaded language, insinuating a claim without actually making it (e.g., by quoting someone else making it), strategically ordering the presentation of facts, and even simply changing the headline can all manipulate the reader's (or viewer's) impression without their awareness." In an age where bot farms are capable of manufacturing a reality that overwhelms the voices of real people, it also leaves us vulnerable to the weaponization of social norms and coordinated attacks from trolls. Creating the infrastructure and partnerships to understand and address this problem at scale is an enormous frontier for researchers, industry, and governments; individuals cannot be expected to take these issues on alone.

But what about us as individuals? What are the ways that we can protect ourselves from being led astray while letting in the useful information that can help us become the people we want to be? How do we learn from other people's triumphs and missteps and reap the benefits of social connection that come from caring deeply about what others think and do, without falling prey to the perils of malicious influence, false information, and bad examples? As neuroscience has shown us, social influence isn't just something that happens outside of us. It depends on what and whom we pay attention to and spend time with, and it fundamentally shapes how we calculate value, how we see ourselves, and who we become.

Understanding how the brain's value calculation works, how we determine what is self-relevant and socially relevant, highlights how subjective these processes truly are and how much the cultures we are part of play a role in shaping our views and behaviors. It also highlights some of the ways that societal expectations and resources can either open up or constrain the possibilities we might naturally imagine for ourselves. Finally, and importantly, it suggests that our decisions can be malleable and subject to both visible and invisible influences. With this in mind, next we'll explore how we can more purposefully curate some of what goes into our value calculations, how to let useful information in, and how to set up our environments to support our bigger-picture goals.

# PART 2

# CHANGE

# 4

# To Change What You Think, Change What You Think About

ERNIE GRUNFELD'S DECADES-LONG CAREER in the National Basketball Association (NBA) is the stuff of a best-selling novel or made-for-TV movie. When he first arrived in New York City in 1964, at eight years of age, Ernie had never played basketball. Just twelve years later, he won an Olympic gold medal for the United States and went on to play for the Milwaukee Bucks, the Kansas City Kings, and the New York Knicks—becoming the only child of Holocaust survivors to play in America's premier basketball league. And yet, his exceptional career might not ever have happened if it weren't for a series of choices that unfolded between Ernie, his parents, and his high school basketball coach.

Before any of this, Ernie's parents, Alex and Livia, were Jews living under the Soviet regime in Romania. For years, they stashed money in an old radio to keep it secret from police searches until they finally escaped to New York, found a home in Queens, and opened a fabric store in the Bronx. The family's routine revolved around the store: Alex worked there seven days a week, while Livia worked six, reserving one day each week to cook, clean, and get the family organized.

On Saturdays, Ernie, too, took the E train, to the D train, to the fabric store, to help with the weekend rush of customers.

On weekdays after school, though, while Alex and Livia worked their long hours, Ernie started playing basketball—first at the playgrounds in his neighborhood in Queens, then on teams at school. By the time Ernie was a freshman at Forest Hills High School, he was six foot two and had developed a reputation as an excellent basketball player. In his sophomore year, he averaged over 17 points per game. By his junior year, at six foot four, Ernie was up to more than 20 points per game and was among the best high school players in the city.

Although Alex and Livia knew Ernie played basketball while they worked at the store, they had no idea how good he had become. Imagine their surprise when Livia picked up the phone one night and heard Ernie's high school coach, Irwin Isser, on the other end of the line. "Mrs. Grunfeld," the coach said, "you need to come watch your son play basketball. He's the most determined player I've ever coached. He'll be able to go to college for this. He's gifted."

There are many ways Livia and Alex might have reacted in this moment. They had organized their lives around not just running the family store but making sure Ernie would be well educated. They paid more to live in Queens—and lived farther from the store—so that Ernie could get a better education, not so he could play a sport. They could be forgiven if they worried that the time and energy Ernie was evidently spending on basketball might take away from his studies or his ability to help out at the store, where they still worked virtually around the clock to make ends meet.

But Coach Isser's call had been well framed. The coach highlighted that Ernie was *the most determined player* he'd ever coached: the Grunfelds prized hard work and dedication. Ernie could be recruited to play basketball in college, the coach said. Far from conflicting with Alex and Livia's educational goals for their son, basketball could open new opportunities for him.

So a few days later, on a cold and windy afternoon, Alex and Livia closed the store early—something they never did—to go watch Ernie's team face off against Bronx High School of Science. Livia recalls seeing her son on the court in the hot, packed gymnasium, but her husband didn't even recognize him in his uniform. Alex lamented that although Ernie was the team captain, he wasn't even playing.

"What are you talking about?" Livia replied. "He's right there!"

After this, the family's routine changed. Alex began attending all of Ernie's games, and Ernie was allowed to stop working at the store on weekends to focus on basketball. During his senior year—just as his coach had promised—Ernie was recruited to play at the University of Tennessee, where he would set a new record as the school's all-time leading scorer. (Nearly fifty years later, Ernie remains second all-time.) He was on his way to a career in the NBA.

But what if Coach Isser hadn't called? Or what if he had chosen different things to highlight during that short phone call with Livia? Whether deliberate or just a lucky accident, by praising Ernie's determination and predicting that his talent would be a ticket to higher education, the coach drew Alex and Livia's attention to the ways that Ernie's basketball career could be consistent with the family's identity, goals, and values, thereby guiding their ultimate decision. Put another way—although he couldn't have thought of it in precisely these terms at the time—the coach focused Alex and Livia's attention on dimensions of the choice that felt highly self-relevant, and ultimately helped tip the scales of their value calculation.

As we learned in the first part of this book, our value system guides our choices according to our past experiences, current needs, and future goals, compressing the many factors of a choice down to a common scale that allows us to compare them. This means that the brain is capable of imagining nearly limitless possibilities. So how does it sort through and prioritize all these elements—and how can we use our understanding of it to help bring our choices into greater alignment with our goals?

As we'll see, the value system puts the most weight on immediate rewards—immediate self-relevance and immediate social relevance: in other words, how relevant each option is to me and people I care about, *right now*. But, critically, the outcome of the calculation is not fixed. The way a message is framed or a situation is set up can shape which dimensions we think of as most relevant to us and hence what we think, how we feel, and what we do in response.

Just as Coach Isser changed the way that Livia and Alex thought about Ernie's future—and, earlier, how Jenny Radcliffe influenced the security guard by focusing his attention on the social consequences of the scene she was making—we can deliberately change what is front-of-mind in our and others' value calculations and thus affect the decisions that follow. It's a way of shining the flashlight on elements of a choice that are most likely to cause someone (including yourself) to consider that choice a little differently. Think of me riding my bike to my grandma's; once *How to Save a Planet* reporter Kendra Pierre-Louis helped me remember how *fun* riding a bike could be, it was easier for me to make a choice that also aligned with many of my other priorities.

Deliberately refocusing our attention can shape choices in lots of other ways as well: doctors can spend more time highlighting the risks or benefits of a procedure; salespeople can highlight the utilitarian features or social cachet of a product; parents can highlight the tastiness of what's for dinner; the user interface designer of a fitness app can choose to foreground your daily steps, calories burned, or minutes of vigorous activity. Coach Isser chose to highlight Ernie's dedication.

When Isser referred to Ernie as the most dedicated player he'd ever coached, Ernie's parents thought of the happiness, academic opportunity, and financial future that basketball could give Ernie, and this convinced them to support his budding career. But when Ernie himself reflects on what drew him to the court every day, it wasn't the prospect of a college scholarship or dreams of a career in the NBA—he didn't even know that was a possibility when he first got started. His motivations were more immediate—and immediate motivations are some of the most powerful.

## WHAT'S IN IT FOR ME, *NOW*?

At school Ernie struggled to fit in. He was teased for eating *ciorba* and *uborkasaláta* instead of hot dogs and pizza and for speaking Hungarian. But at the park it was different. There, Ernie fit in with the other basketball players. He started to pick up English. Most of all, he had *fun*.

A key function of the value system is to help close the gap between where we are and where we want to be by guiding us toward the options it predicts will be most rewarding and then continually updating those predictions, depending on what actually happens. Yet, as we've also seen, the value calculation is shaped powerfully by the situation we're immediately in—especially what we judge to be most self-relevant and socially relevant in that moment. This present bias can be good (it can help us stay focused in the moment) but it can also mean we get sidetracked by moment-to-moment influences that might be less compatible with what we imagine to be our longer-term goals.

The quirks of how self-relevance is calculated are particularly important. As seamlessly as we're able to mentally project ourselves into different times and situations, our value systems tend to prioritize the here and now—and this tendency is even more powerful than you might imagine. When people are asked to think of themselves in a different time or place, their brains react in much the same way as when they imagine a different person entirely.

In one study led by Diana Tamir with Jason Mitchell at Harvard, volunteers were asked a variety of questions about themselves (for instance, "How nervous would you feel speaking in public?") and then asked the same question with different parameters. How would they feel now, or in a year? Here, or in Oxford? As themselves, or a specific other person, like Barack Obama? With their own identity, or another (for example, as a woman or as a man)? Diana's team found that activity in the volunteers' self-relevance system changed, corresponding to the distance they were imagining themselves across, regardless of whether they were traveling in time, space, or identity. Activity was highest for the self who was right there, right then, with the identities the volunteer held, and lower for imagined selves who were farther away geographically, temporally, or in identity. In other words, imagining themselves in the distant future affected the self-relevance system similarly to imagining themselves as someone like Barack Obama or in a faraway city like Oxford. "Me" in another time or place or body was akin to another person.

You can imagine how this would affect the value system, which we know is closely entwined with the self-relevance system. If Future You is more like a different person to your brain's value system, then benefits accruing to Future You are seen as less self-relevant, and therefore less valuable, than ones you would enjoy here and now. Thus, the calculation tends to weigh the Current You more heavily. It's why you might reach for a delicious cookie even when you're trying to eat healthier, or go out to a party with friends instead of plugging away at a project for school or work. Chances are you know that staying in and working would be useful for your upcoming test or deadline, as well as your longer-term career goals. You likely also know that having a few drinks might make you feel worse in the morning. But, hey, that's Future Me—they can deal with that issue later! Right now, *I'm* having fun.

Psychologists call this tendency to value rewards here and now *temporal discounting*, or *present bias*, and my University of Pennsylvania colleague Joe Kable ran a series of clever experiments that homed in on just how this works in the brain. He and his team had volunteers choose between receiving $20 now or a larger reward (between $20.25 and $110) at different times in the future, from six hours to six months from now. If you were a volunteer in Joe's study, you'd be asked, for instance: Would you rather have $20 now or $21 tomorrow? Twenty dollars now or $21 in six months? Twenty dollars now or $100 in a week? As you might expect, people naturally vary in how patient they are—some are willing to wait a long time for only a little more money, while others require a lot more money to be willing to wait even a short time. But across the group of volunteers, the less additional money they were offered, and the longer they had to wait, the more likely the volunteers were to just take the $20 now.

The researchers also saw less activity in key parts of the volunteers' value (and overlapping self-relevance) systems in response to these relatively lower or farther-off future offers. The subjective value of later rewards was discounted in their minds—it seemed less valuable to them than sooner rewards. Later research also showed that the tendency to be patient and wait for future rewards seems to be related to how closely people's representations of their current selves aligned with their future selves in their brains. The more similarly people's brains represented their current and future selves, the more willing they were to give up rewards now in favor of later benefits.

If you've ever bought something you wanted now instead of saving a bit for retirement, or gone out now instead of working late, then it probably feels pretty obvious that it takes more effort and energy to pursue goals where all the benefits appear to lie far off. Yet we frequently fail to take this tendency into account when we decide to work toward a goal. When we try to motivate ourselves to change—for example, to eat healthier, be more patient with family members, or develop a new skill at work—we most often focus on long-term, relatively abstract benefits and consequences (I want to live to be one hundred years old!). In other words, the way we overwhelmingly attempt to inspire ourselves and others to change is fundamentally at odds with our brains' default priorities. So how do we bring current and future rewards more into alignment?

In 2019, a team of researchers at Stanford University ran an experiment in five university dining halls to explore that question. Their study showed that emphasizing the (immediate) tastiness of vegetable dishes, with dining hall labels like "Herb 'n' Honey Balsamic Glazed Turnips" and "Sizzlin' Szechuan Green Beans with Toasted Garlic," was significantly more effective at getting college students to choose vegetable dishes than emphasizing how healthy they were ("Healthy Choice Turnips," "Nutritious Green Beans"). Across six months, twenty-four types of vegetables, and nearly 138,000 dining decisions, people chose veggies with an appetizing label promising (immediate) tasty rewards right away 29 percent more often than those with labels touting (long-term) health benefits. A later study by the same team yielded even more impressive results: emphasizing the tastiness of various healthy options in different dining locations led to a 38 percent increase in healthy dining choices like vegetables and salad. In each case, working with people's natural tendency to prioritize immediate rewards helped them make healthier choices—even more so than focusing their attention on longer-term health benefits.

In addition to reconsidering how we label and talk about our own food choices (or food options for people we love), we can put these findings to use in other places—we just have to look for the immediate rewards in the things that feel effortful (but that we know will be good for us in the long run). If you hate networking because the future career benefits seem abstract compared with the immediate awkwardness of making small talk, think about one or two colleagues you do actually like spending time with, and start by spending time joking with them, rather than focusing on how important it is for achieving your future career goals. If you have kids you're trying to encourage to read more, consider letting them pick books that they *enjoy* most, rather than books that are most likely to optimize their future academic success (same goes for adults looking to trade good old-fashioned reading for doomscrolling before bed). If you're trying to learn a musical instrument, choose songs that you really enjoy playing to start your practice sessions.

Sometimes it's hard to find a way that the behavior you want to adopt—the one that benefits you in the long run—is also rewarding now. In these cases, it can be helpful to pair that activity with another you do find immediately enjoyable. If you can't find a colleague you want to joke with at the networking event, can you eat a delicious dessert as a reward? When one of my colleagues at the University of Pennsylvania, behavioral scientist Katy Milkman, was in graduate school, she found it a slog to go to the gym. No amount of endorphins could provide enough short-term motivation for her. At the same time, she was finding it difficult to resist her desire to listen to her favorite fantasy fiction series at the times when she was supposed to be studying for her exams. What's a grad student to do?

Katy decided to reserve doing what she *wanted* (to listen to exciting fantasy books) for times when she was trying to do the thing she struggled to do (get to the gym). And it worked. She was less distracted from studying, and going to the gym offered a treat. The hard thing was more immediately rewarding and became easier to do.

Katy later tested whether this kind of *"temptation bundling,"* as she named her technique, could help others get more exercise as well. She and her team gave one group of research volunteers an iPod preloaded with their favorite audiobooks, which they could access only at the gym. Meanwhile, members of a second group were given gift cards to buy audiobooks with no workout strings attached. It may come as no surprise that the first group worked out more, but how much more is really striking: they exercised 51 percent more frequently than the second group.* The long-term value of exercise was the same for both groups, but temptation bundling changed the value those volunteers experienced *now*—and that's a profound way to make a choice easier.

Although the effects of temptation bundling declined over time in Katy's original study (particularly following Thanksgiving), temp-

---

* Impressively, a third group that was encouraged to limit their audiobook listening to the gym (but not prevented from listening elsewhere) also increased the frequency of their gym attendance—by 29 percent relative to the group that had totally free rein.

tation bundling can be a way to get over the barrier to getting started and overcome the present bias.

In this way, short-term thinking can become an asset that helps us both achieve our goals for the future and find more pleasure in our present. We can close the gap between current pleasures and future outcomes by setting our focus on ways to make the behaviors that are consistent with our long-term goals more rewarding now. We can refocus our attention on different aspects of a situation or choice to overcome present bias (for example, playing basketball is fun and lets me bond with friends now, rather than a way to earn a scholarship; networking is a way to connect with a few colleagues and eat a free lunch, rather than a way to maximize later career success) or add something fun to an unpleasant thing (listening to *The Hunger Games* at the gym). We can highlight "what's in it for me, *now*." But another way of changing your focus in the moment is to actually connect with the person you will be in the future.

## MEETING FUTURE YOU

What would you do if you came face to face with your future self? After you got over the shock of how much you look like your parents, would it change how you behaved in the present? That's what Stanford researchers Hal Hershfield, Jeremy Bailenson, and their team sought to learn when they put volunteers in a virtual reality situation and gave them the chance to interact with an aged avatar of themselves. And it very much did change how they behaved in the present. Afterward, these volunteers allocated twice as much money to their retirement fund.

This is a remarkable change when you think about it, especially since, by default, our value systems focus on the things that are most immediately salient—the implications for the me that is right here, right now. In fact, the experiment was so successful that Bank of

America/Merrill Lynch created a tool called "Face Retirement," in which customers were shown aged versions of their own photograph to nudge them toward saving more for the future. But you don't need a virtual reality interaction or even to see an aged picture of your own face to get in touch with Future You; you can change some of how your brain weighs the value of different choices just by deliberately deciding to focus on your future self.

We may default to focusing on the now, but we can actively decide to focus on the benefits for our future selves. As we saw earlier, the more people's brains distinguish between "current me" and "future me," the less likely the person is to find the idea of future rewards rewarding and patiently wait for them. So how can we work on closing that gap for ourselves or for the people we care about?

Neuroscientist Hedy Kober explored this question to help people reduce their cravings for cigarettes and junk food. In several studies her lab conducted, smokers looked at photos of the delicious junk foods or of people smoking. For each picture, the person was instructed to think about the immediate (typically pleasurable) consequences (for example, how would it feel to smoke the cigarette now?), or to think about the long-term (typically negative) consequences of consuming the substance over time (for example, what consequences would be likely if you kept smoking for the next few decades?). Nothing else changed—just what the participants were *thinking* about—and yet, their value systems responded differently. Thinking about the long-term consequences dampened activation in key parts of the value system, which correlated with dampened cravings as well. Likewise, in another study, although people's default tendency was to prioritize how tasty different foods are when deciding what to eat, when researchers prompted them to pay attention to different reasons they might choose to eat one food or another—that is, how tasty it is or how healthy it is—it changed the way their brains valued the foods and the choices they ultimately made. Coaching people to focus on health made them about twice as likely to choose healthy, but untasty, foods and also reduced their tendency to choose tasty, but unhealthy, foods. In other words, it tipped the scales in favor of the healthy options.

Shifting focus away from the immediate rewards to the longer-term consequences dampens the value signal and cravings that lead to the less healthy choice. This relatively simple strategy—taking a minute to think about the personal, longer-term consequences of behaviors—is commonly used by therapists helping people reduce their substance use. But it's also within our personal control to do this. Brain imaging research suggests that people are capable of engaging in this process without being explicitly instructed to and that it can impact cravings and behavior. If the flashlight typically shines on the present, we can redirect that beam to the future—and not just future *you*, but the future you want to live in.

## A CLEAR PICTURE OF THE DESIRED FUTURE

When Livia heard from Coach Isser, she may not have known much about basketball, but she did have a very clear priority that Ernie get a good education—she wanted him to have what she did not have access to during the war. This meant that when Isser suggested that basketball could be Ernie's ticket to college, he was already tapping into a desired future that was clear and important to Livia. It may have helped her to quickly pivot her present expectations and choose to support an alternate path for her son.

Having this kind of clarity about a desired future can help many of us do what Livia did: move forward in an unexpected direction that seems to lead to the future we want. We can do that in big, grandiose ways, and we can also do it in smaller, practical ways.

Choosing a new team member is arguably one of the highest-stakes decisions we make at work, with ramifications that can last for years. It requires thinking about, and often trading off between, a variety of skills and qualities different candidates might possess. In my lab, for example, we routinely hire research coordinators who need to be, at once, great with people, great at data analysis, highly organized, able to work independently, able to collaborate with people across the university, local community, and more. We often get hundreds of applicants for each opening. Ideally, at least one of them will tick all the boxes, but most people will be relatively stronger in some areas than others. So how can we tackle these hundreds of applications in a way that will find the person who is best suited for the job?

Just as we saw in Livia, it helps to have a clear idea of the kind of future we want—in this case, what attributes would be most helpful in a future colleague in this position? Harvard psychologists Linda Chang and Mina Cikara have studied how the attributes people pay attention to when hiring make a big difference—and also how easily people can get sidetracked or misled during the decision process. Imagine that you're interviewing a series of candidates for a job, and the first person you interview is very awkward, though they have other useful skills. Suddenly, interpersonal skills take on heightened importance (even though other attributes might ultimately be more important for doing this job well). In your next interviews, you might overweight the extent to which candidates seem friendly and charismatic, because the first candidate made these characteristics stand out. Although interpersonal skills might be generally desirable in a colleague, and more fun during the interview, this might cause you to overlook other relevant dimensions, like how much programming experience the candidate has, that are more important for doing the job well over the longer term (darn present bias!).

To test this, Linda and Mina created a situation in the lab where people were given a few pieces of information about different fictional job candidates. Even in this simplified context, the volunteers changed their preferences when different attributes were made more salient. In particular, when the research team introduced candidates into the hiring pool who were inferior on one specific job-relevant dimension (think social skills), it made that dimension stand out to the volunteers and increased their preference for a candidate who had the skill the other lacked, even if the second candidate wasn't the best on other dimensions. And it's not only attributes but identity that can skew a hiring process—and not always in the ways we imagine.

Knowing what we know about the self-relevance and social relevance systems, it is not surprising that people value traits that they themselves have over other measures of competence. For instance, founders of US startups are often more likely to hire employees who are like them than people who bring diverse skills to the table. To offset this bias, Linda and Mina's results highlighted that structuring hiring decisions so that people directly compare candidates on predefined criteria weighted in a predefined way (rather than across a range of different criteria defined on the fly) can make people less biased and reduce errors in choosing the right candidates (according to the predefined criteria). Of course, this means that the predefined criteria matter a lot—and we need to be very aware of our biases as we define them.

In addition to getting clear about what kinds of experience and expertise we want to prioritize in our hiring, we can also shift how we evaluate experience. For instance, a meta-analysis of over one hundred studies on more than ten thousand teams highlighted that greater diversity in teams is associated with better reasoning, more creativity and thoughtfulness, and idea novelty. Indeed, many organizations would like to diversify their teams, but then find themselves hiring the same sorts of people. There are many reasons why this might happen, including long histories of power imbalance and inequality across groups. This also shows up in the ways that we think about what we value in a prospective recruit and what constitutes relevant experience for the job. We need not just a clear picture of the future—what's *really* important to us—but also imagination in how we get there.

When reviewing hundreds of applications, it's easy to default to looking for candidates with experience that resembles ours or our colleagues'—self-relevance can be a trap. Remember the ways that self-relevance and value are intertwined, making it more immediately comfortable for us to interact with someone "like me" who I might think would easily fit into the existing company culture. But opportunities to gain experience—for example, through unpaid internships or social connections to people who can offer first jobs—are not equally available to everyone, especially to people from marginalized groups. If we don't correct for this in the hiring process, we can miss out on great people.

Rather than focus narrowly on applicants who have worked with research participants in a lab before, my team might consider how working as a server in a restaurant could make someone great at interacting with research participants, too. Likewise, going back to the first candidate who seemed "awkward," when we interview someone whose perspective is different from ours or who expresses their opinions differently than we do, we could think of them as someone who is brave enough to share new ways of thinking ("exactly the right person for the job"). By focusing on a wider range of experiences that could make someone a good fit, we are more likely to hire a diverse team of superstars. Reframing what you're thinking about from "experience that looks most like mine" to "experience that complements mine" or "a team member who will bring new perspectives" can change what you value in a candidate and therefore whom you hire.

So changing focus to highlight a particular dimension of a choice can influence what we buy, what we eat, whom we hire, and how we invest our money. Highlighting how a particular choice aligns with our future can change how we choose, but like we saw in the hiring process, it's not just the clear picture of a future, but a flexibility in how we imagine that future might take shape, that allows us to benefit from new and different ways of thinking. It's about *changing* our focus and where we direct our attention.

Indeed, where we focus our attention is related to what we value and how we choose. For example, scientists have found that drawing attention toward or away from price alters how people value different products. If you first learn the price of a product before finding out other things about it (placing more focus on whether it is "worth it"), it will shift your attention and brain's value calculation to how practical the item is, rather than how much fun the product is. But if you learn the price after falling in love with the product, practicality is weighed less. This tendency can be used by advertisers, but it might also help you be more persuasive. If you're talking to a family member about where to allocate your shared budget, and you want to nudge them toward a practical kitchen appliance, you might mention the prices of the different options before their features, shifting the focus there. But if you want to nudge them toward a guilty pleasure (not that you would ever do that), highlighting the ways the product would bring joy before mentioning the price is likely to be more effective.

And it's not just when, but *how long* we linger on a particular element. The more time we spend focusing on a particular option, the more likely we are to choose it, or convince someone else to choose it. In one study, people were given choices about whether to take a gamble in which they had an equal chance of winning or losing different amounts of money. As they considered their choices, researchers tracked where they looked. When people spent more time focused on the potential loss, they were less likely to take the gamble, even when they stood to win money. So our attention is correlated with our preferences. In another study, researchers found that making a stock's original purchase price less salient to investors helped reduce their biased tendency to sell at unfavorable times. In other words, redirecting attention can also change how we make choices.

So far, we've covered a number of ways to focus on different elements of a choice in order to change our perspective or even to potentially bring our choices more into alignment with our goals— for instance, we can decide what dimensions of a choice we want to give most weight and spend more time and focus on those. But sometimes we don't get to choose. Sometimes choices are made for us.

And though this can be a struggle when it happens, it turns out that similar tools can also help us change how we *feel* about a situation as it unfolds—even if the situation itself isn't under our control.

## TO CHANGE THE WAY YOU'RE FEELING, CHANGE THE WAY YOU'RE THINKING

The life of a professional athlete, however privileged, comes with certain constraints. For example, although some of us have some ability to choose where we live and the organization for which we work, an NBA player has little control over this, especially early in their career. True to Coach Isser's prediction, Ernie had had his pick of colleges, ultimately choosing Tennessee—but when it was time to enter the NBA draft a few years later, his destination was largely out of his hands. "The way the system works you don't really get to pick where you go," he explained to me. "They pick you and you have to go there."

Having performed exceptionally well in college ball, Ernie was likely to be a first-round pick. That year, Ernie's hometown team, the New York Knicks, had the 10th pick in the draft, followed by the Milwaukee Bucks at 11 and the Boston Celtics at 12. Ernie desperately hoped the Knicks would pick him ("Of course, everybody always wants to play for their hometown team"), but Boston would be great also—it was just a train ride from New York, and the Celtics coach had indicated interest. When draft day came and the Knicks selected a friend of Ernie's, he figured he'd end up in Boston. But in the next round, he learned he'd be headed to Wisconsin. Ernie had never been there, and it was far from his family. Understandably, he was disappointed. But when his mother asked him if he was happy about the move, he told her, "I'm going to *make* myself happy." And he did.

"I would have preferred going to New York, of course, and Boston was always a marquee franchise, so that would have been a nice opportunity. But [Milwaukee] ended up being great," he told me. He recognized that even if he didn't have a choice about where he'd kick off his career as a professional basketball player, he did have a choice about how to think about it and therefore how to feel. "It's all about doing what you have to do sometimes—sometimes you don't have choices and sometimes you do have choices." Ernie says he could have been upset, not played well, not listened to the coach, or not bonded with the group. But, said Ernie, "that's not my makeup. My makeup is to go there and make the best of the situation."

Although most of us aren't likely to find ourselves in the exact same position as Ernie (unless we happen to be spectacularly gifted athletically), and you might not feel that bad for a professional basketball player, most of us experience different versions of being in situations where we have little control and don't get our first choice. Maybe you don't get accepted into your dream college as a teenager, or maybe as an adult you don't land your dream job. Sometimes the person you'd most like to date doesn't reciprocate your feelings, and sometimes the movie you brought your date to see is sold out.

When I was upset or frustrated as a kid, my mom would tell me, "If you don't like the way you're feeling, change the way you're thinking." I didn't really get it at the time, but now that I understand more about how the brain works, I see that in many cases she's right.

As we know, the brain's value system doesn't only calculate the expected reward of different decisions; it also keeps track of the actual reward we experience. This is another way of saying that it keeps track of our feelings. Brain imaging research shows that activity in regions of the brain that overlap with key parts of the value system is associated with positive and negative emotions. Specifically, parts of the ventral striatum and medial prefrontal cortex that are core to the value system seem to correlate with how positive or negative we feel. When researchers Amy Winecoff and Scott Huettel, of Duke University and Caltech, showed people pictures that typically evoke positive emotions (think cute puppies, babies, beautiful nature scenes, erotic nudes), activity in their value systems increased. But looking at pictures that tend to evoke negative emotions (think spiders, snakes, wounds, and violence) decreased activation. Importantly, these activity patterns corresponded with how each volunteer reported the pictures made them feel: the more positive they felt, the more activity in the value system the researchers recorded.

If our value system is partly responsible for keeping track of how we feel, this suggests that we can use tools much like those we explored earlier in this chapter to change not only what we choose but how we feel about it. Additional brain systems (involved in what neuroscientists call cognitive control) change what we pay attention to and how the value calculation unfolds. Psychologists have a name for purposefully changing how we think about a situation to change how we feel: *reappraisal*. Research by Kevin Ochsner and his team at Columbia has shown how teaching people to reappraise activates the cognitive control system, and in turn, helps them change how negative they feel about a wide range of challenging emotional experiences. Most often, reappraisal techniques aim to help people dial down negative emotions by turning their attention to different aspects of a situation that make the person feel less negative. In other cases, it can be helpful to teach people to work with, rather than against, uncomfortable feelings like anger or fear, which can drive bigger changes. Sometimes it's broader systems and not individuals that need to change, and anger can motivate actions like getting involved to change things. Sometimes fear can motivate changes that are best for us in the long run. In other words, "liking" the way we are feeling doesn't always mean it feels good—instead, we can work on appreciating feelings that are compatible with our bigger goals and values. And, in the same way that we can focus our attention on different aspects of a choice to make certain options feel closer or more relevant to us, we can use similar techniques to dial emotions up or down, according to our goals.

Ernie really wanted to go to New York to play for the Knicks, but found himself in Milwaukee. Instead of focusing on everything he was missing out on in New York, he focused on the chance to play professional basketball and also on a budding romance with the daughter of an attorney who was key to bringing the Bucks to Milwaukee. But even in much smaller, everyday circumstances, we can reappraise to get more enjoyment out of daily life and be a little easier on ourselves and each other. In *The Book of Delights*, Ross Gay describes how terrified he was watching *The Exorcist* as a kid and how his experience of the film completely changed watching it at age twenty-six with a group of fellow moviegoers who made light of the terrifying scenes, shouting "Oh no she didn't" and "That girl is trippin'!"—which helped him experience silliness rather than fear. Likewise, if someone I care about answers my phone call by asking me why she hasn't heard from me, instead of getting angry ("What do you think I just did? We're on the phone now!"), I'm better off focusing on the fact that she loves and misses me.

Reappraisal can also help us regulate our positive emotions, whether we want to amplify them or dampen them. At first blush, it might be less obvious why the latter would be helpful—why make yourself feel *less* happy or excited? But there are some situations in which getting too carried away by positive feelings can be detrimental. For instance, if you're apartment hunting and fall in love with the bay window in one unit, it might distract you from the fact that the sink leaks and the apartment is close to a noisy train station. Similarly, let's say you get offered your dream job. Your excitement at the opportunity could make you leap to say yes, missing the chance to negotiate a better package for yourself in the long run. In cases like these, exerting some effort to dampen your positive emotions through your cognitive control system could be helpful. You might ask yourself what parts of a situation are going to matter most for your longer-term happiness and focus most heavily on those.

Whether you're trying to control positive emotions or negative emotions, the techniques you can use to change how you feel are similar. One, as we've already seen, involves changing where you are focusing your flashlight in a situation and coming up with new interpretations. In your apartment hunt, if you felt disappointed when someone swooped in and took the place with the bay window, you might say to yourself: "It's okay that I missed out on that apartment because the train noise would have driven me crazy, the extra half hour of commute would have been hard, and negotiating with the landlord to fix the leaky sink would have been annoying." After a rough breakup with the partner whom all your friends loved, you might revisit your parents' concerns about the relationship and think about how your longer-term happiness will benefit from finding someone who wants to live in the same city you do. Leaning into negative feelings about the consequences of behaviors that aren't great for our health, like binge drinking, can also make us more open to following useful health advice.

Another useful form of reappraisal is to try stepping outside the situation, imagining yourself as an objective observer. This was something that Amy Winecoff and Scott Huettel explored in their study of how people's value systems responded to pictures designed to evoke negative or positive emotions. When they asked volunteers in their study to "imagine themselves as an objective observer to the situation depicted" in the picture they were being shown, "or to imagine the event as having no personal relevance to them," both their positive and negative feelings felt less strong than when they first reacted naturally. The value system reflected these changes, with activity increasing when people reappraised negative images more neutrally and decreasing when they dampened their responses to positive images.

This form of distancing might be particularly helpful when the emotions of a situation are likely to skew the way you think about it. In a case like this, you might imagine the situation happening to someone else, imagine yourself as a fly on the wall looking down from a distanced perspective, imagine that the situation is happening far away, or otherwise imagine that the situation is less directly about you. As we'll explore more in the next chapter as well, taking a step back to imagine the situation with some distance (How would an outside observer see this situation? How might I look back on this in five years?) can make our immediate emotions less "hot," increase our wisdom, and improve our decisions.

It is worth noting, though, that reappraisal takes some conscious work. A related technique, simply taking the time to notice how we feel and approaching those feelings in a less judgmental, reactive way, can also have benefits and change key brain responses in the value system. If we take the time to notice how we actually feel *now* (*that generic supermarket birthday cake feels a little . . . gross*), it can help us make choices that are better aligned with our current needs rather than past scripts.

UC Berkeley neuroscientist Hedy Kober has found that teaching people to approach different feelings in this more mindful way can have benefits for their health and well-being. For example, smokers with no meditation experience were able to reduce their neural and self-reported cravings when they learned to mindfully attend to smoking-related images. In another study, smokers who went through mindfulness training reduced their smoking in both self-reported and objective measures of behavior even seventeen weeks after the training. Research suggests that this may work by lowering the value system's response to smoking-related stimuli, as well as reducing certain brain regions' reactivity to stress. In other words, being more mindful about where we direct our attention can alter how our brains compute value and reward.

If you want to change the value of different choices for yourself or someone else, changing how you think about the situation can change how you feel, which can in turn change what you do. It worked for Ernie. When he told his mother he was going to make himself happy in Milwaukee, he was right. Not only did he enjoy his time playing basketball for the Bucks—he also met and fell in love with his future wife there. And in the long run, not only did he have an illustrious career as a player, but he also served as president and general manager of the Bucks and the Washington Wizards, and yes, eventually he did end up back in New York—first as a player and later as the general manager of the Knicks.

Ernie's story highlights the luck we have when the things that we love and that bring us immediate feelings of reward turn out to be compatible with our bigger-picture goals. Ernie loved playing basketball, and he was lucky enough to be incredibly good at it. But what happens when the things that seem immediately pleasurable in the short term don't obviously match up with our bigger-picture goals? Here, we have a different kind of choice. We can choose to just appreciate the pleasure of eating a delicious piece of cake and not worry about the longer-term health consequences—or we can deliberately focus on the longer-term consequences of eating the cake or what we like about a healthier alternative. Changing what aspects of a choice or situation we shine light on can change what we choose and how we feel. We can also draw other people's attention to different aspects of a choice and shape their decisions, as Coach Isser did when he highlighted what Ernie stood to gain from devoting his life to basketball.

But it isn't always easy to take coaching. We get defensive and come up with reasons why the paths others suggest might not be right for us. As we'll explore more in the next chapter, another part of Ernie's philosophy is also instructive here. Throughout his life, whether on the court or off, he focused on connections with other people—a desire to connect with other kids on the playground, the love and admiration he felt for his parents, the warmth he felt for his teammates and the players he managed, his excitement to spend time with his grandchildren. In the next chapter, we'll explore more deeply what might happen if the rest of us let go of our fixed sense of self and broadened our focus more to connections with those around us.

# 5

# Become the Least Defensive Person You Know

MY PARTNER, BRETT, SAVES up stories all day. After our kids go to bed, in the hour or so before the two of us turn in, we often hang out in our kitchen, and he'll tell me about interesting things he's read or funny things that happened to him that day. "Do you want to hear about what the Internet says kids tell their therapists about their parents?" he'll ask, or "Faculty meeting today was wild!" I find this habit endearing, and I love this hour of the day when we get to reconnect.

On one recent evening, while Brett finished doing the dishes, I was snuggled up under my favorite soft pink blanket in the plush armchair at the end of our dining table, responding to an email from a student who needed my approval to move ahead with their research—in short, enjoying one of my favorite moments of the day. When I heard Brett's voice, I first thought he was about to share one of his stories, but then I noticed his tone: hesitant and serious.

"Do you think you could leave your phone in the other room while we hang out?" he asked. When I'm on my phone, he told me, it makes him feel like I don't think his stories are interesting, and it makes him want to check out, too.

"I'm paying attention to what you're saying," I assured him, still looking at my phone. "I just need to deal with these few emails." If I then went on to check Twitter, maybe I'd see something there that he'd find interesting too. Sometimes I was looking up information, like statistics, directly relevant to our conversation, I pointed out.

When I looked up from my phone, Brett was looking directly at me, one eyebrow slightly raised, the New England equivalent of a deep, exhausted sigh. It turns out, he wasn't impressed with my multi-tasking skills, and he didn't find that the stuff I looked up on the fly added to our conversation.

Of course, Brett was right, and I knew it. Most of us hate it when the person we're trying to talk to is on their phone. Experiments run by psychologists have shown that phone use by one party—in fact, even the presence of a phone, whether on a table or stashed out of sight—actively decreases the quality of interactions between people. The same research shows that people who report that other people's phone use makes their time together worse still try to justify their own, the same way I did. I can remember times when I have felt annoyed by a friend texting while we were on a walk or by a family member texting at dinner, and yet it was difficult for me to admit that I was making Brett's and my time together worse. But then I thought about how enjoyable it is when we do give each other our full attention and how much I appreciate friends who never check their phones when we're together.

Over the next few days, I chewed on it. I experimented with leaving my phone upstairs after the kids went to bed, removing the temptation to check it while Brett and I were in the kitchen together. Brett said he noticed that I was using the phone less, and that felt good. Still, I'll admit, it was hard for me to step back in the first place and recognize that Brett's request to leave the phone behind was right.

Brett tells me that he had brought the issue up before. Apparently, I didn't hear the request clearly. It's possible that I was on my phone the first few times he asked, but it's also possible that I just didn't want to hear what he was saying. We often don't notice when the psychological systems that defend our sense of competence and self-worth are at work. As we explored in Chapter 2, in many cases the brain systems that assess the self-relevance and value of possible choices substantially overlap, meaning that we are wired to conflate self (our sense of who we are) and value (what we perceive as good). This tendency helps protect our self-esteem, but when we're confronted with prompts to change, it can translate to a defensiveness that reflexively shuts down information that we could benefit from.

If you've ever tried to convince a loved one to change their mind or change a habit—or if you've been the recipient of unsolicited advice yourself—you can probably relate. Maybe you suggested to a family member that they study harder, exercise more, or be more patient. What reaction did you get? We could all work more diligently, be more active, or take a few deep breaths before responding to an aggravating situation, but most people don't appreciate having this pointed out to them. We want to feel good about ourselves, and messages that suggest we aren't behaving optimally can threaten our self-image. We get defensive and come up with reasons why the advice doesn't apply to us. At the neural level, we can see activation in neural alarm systems go up, and activity in self-relevance and value systems goes down when people react defensively to messages pushing them to change. Our tendency to conflate self and value leads the brain to see *me-as-I-have-been-behaving-or-thinking* as good and to discount the message as not relevant to that version of *me*. In counterarguing these messages that feel threatening to our identity, we double down on our old habits and resist change that could be good for us.

Compounding the problem, our defenses are most likely to come into play when the stakes, and potential benefits to us, are highest—when the topic really matters to us or is part of our core identity. The stronger our habits or prior beliefs are—in other words, the more central to *me* they feel—the more likely we are to get defensive when they're challenged. I care deeply about being a good partner, so when Brett told me it hurts him when I use my phone while we're together, at first I couldn't hear it, and then I really wanted to find a good justification for my behavior. But sometimes the values that make us defensive can also propel us to make different choices.

It's important to recognize, of course, that not all advice we get is good advice or the right fit for our particular circumstances and personal goals (totally fine to ignore the person who suggests you "key your car so that people think you're cool enough to have enemies"). But when we let our defensiveness get the better of us and too reflexively define new information and ideas as "irrelevant," we miss out on new perspectives that could be useful. Leaning into the possibilities that these perspectives may open up for us could allow us to have more productive conversations about justice in the workplace or in our communities more broadly, to communicate across political divides, or to help us work on being a better friend, boss, or teammate.

Understanding how the overlap of self and value in the brain gives rise to defensiveness—and how the self-relevance, social-relevance, and value systems work together to make the value calculation—gives us tools to lessen the impact of defensiveness and makes it easier to find value in new ideas and behaviors. When we see new possibilities through the eyes of others, we have the option to make changes and move forward in new ways. But doing so is far from easy—our egos are well trained to protect us.

## MY MUG

Our tendency to conflate what is *me* and what is *valuable* leads us to hang on to all sorts of things that might be better to let go of. When I was in college, for example, some friends and I once took a bus to Newport, Rhode Island, and spent a day wandering the quaint brick- and cobblestone-lined streets and exploring the cliff-side mansions hugged by stone paths and green shrubs, overlooking the sea. Down one side street, we came across a box labeled "free" that contained various treasures. I took a mug that someone had evidently painted at one of those paint-your-own pottery studios. The artist had sloppily rendered a boat and some water, along with lobsters singing (as indicated by little musical notes emitting from their little lobster heads). I brought the mug home as a joke. On more than one occasion in the decades since, Brett has wielded his slightly raised eyebrow to delicately nudge me to throw it away. (Monstrous, right?)

"No way," I say every time. "That's *my* mug."

Maybe, like me, you have a cupboard in your kitchen that contains random mugs and other clutter. I wish I could tell you that my Newport souvenir is the only mug I've held onto. Some of my mugs I got as conference swag or on other trips I've taken; one particularly prized mug, created by cognitive neuroscientist Talia Konkle, commemorates our friend and fellow cognitive neuroscientist Marina Bedny's sage advice about standing up for things you care about: "When I blow my top, I only ever regret it in the short term." Some of the mugs have been around a long time, and I have no memory of where they came from. But, as I tell Brett, they are all *mine*. This is small comfort to Brett, who hates clutter.

But the data show I'm not alone in my stubborn unwillingness to surrender my mugs. Amazingly, this mug phenomenon has been the subject of study by not one but two Nobel Prize–winning researchers. Psychologist Daniel Kahneman and economists Richard Thaler and Jack Knetsch published a famous mug-based experiment in 1990, in which they found that people gifted a mug quickly come to think of it as their own and will effectively give up money to keep it. In a class at Cornell, students in alternating seats were given Cornell coffee mugs. The students were told to examine the mug they had been given or, if they didn't get one, their neighbor's mug. They were then given instructions, which varied depending on whether the student was in active possession of a mug or sitting next to someone with a mug. The students with mugs were told: "You now own the object in your possession." They also had the option of naming a price for the mug, to attempt to sell it to a peer, or they could take it home with them. The students sitting next to them were given parallel instructions: "You do not own the object that you see in the possession of some of your neighbors." They were told that they would have the option of buying the mug, if the price named was appealing enough to them, or leaving the experiment without paying anything, but also without a mug.

When given a mug and offered the choice to sell it, people demanded twice as much money for "their" mug as people were willing to pay when given the option to buy a new mug. Objectively, both sets of people (the buyers and the sellers) were in the same financial situation—choosing between a mug and money—but the financial value they gave the mug differed significantly. Sellers who felt ownership of the mug (because they had been gifted it at the start of the experiment) were actually willing to give up money they could have earned from selling it, just to keep the object ("it's mine!").

Despite my personal affection for mugs, this isn't a magical property of ceramic drinking vessels. Kahneman, Knetsch, and Thaler found that the same pattern held true for people initially given pens or chocolate and given the opportunity to sell or trade these items. Even in this situation, people tended to hold on to "their" pen—or whatever object they received first—much like the previous participants held on to "their" mug, indicating that this interest in holding on to an object isn't a property of the object being traded but instead the sense of ownership attached to it.

People are also reluctant to make trades that might increase their happiness if they are initially given an object and therefore think of it as their own. The researchers saw this play out in another study, in which students in three different classes were given a prize for completing a questionnaire. In one classroom, all students were given a mug at the start of class and then, at the end of class, offered the chance to trade it for a chocolate bar. In another classroom, all students were given a chocolate bar and then, at the end of class, offered the chance to trade it for a mug. And in a third classroom, all the students were given the choice between a mug and a chocolate bar at the start of class. In the class where people initially received mugs, 89 percent kept them, compared with 56 percent in the class where they had the choice up front. In the class given chocolate bars at the start, 90 percent of the students hung on to them. And these were only objects people had owned for a few minutes! No wonder it's hard to get rid of the keepsakes that have spent years attaching themselves to us with their little lobster claws.

This tendency to cling to things we think of as ours, or that are associated with our identity, is called the *endowment effect*. Neuroscientists have found that key parts of the brain's self and value systems respond differently when people are given an object and then offered cash to sell it, compared with when they're faced with simply choosing between the object and cash to begin with. Again, these choices are functionally the same, since selling the object inherently involves choosing between having the object or getting a given amount of cash. But different parts of the brain see these choices differently.

Activation in parts of the value system, like the ventral striatum, correlated with how much people liked the products regardless of the way their choice was framed (whether they were potentially giving up their object or had the opportunity to purchase the object), which you might expect to be what drives people's choices. But the researchers saw something different when it came to activation in the medial prefrontal cortex, a part of the value system that integrates decision inputs and processes self-relevance. This part of the brain was susceptible to the ownership framing, which might help explain why I would want to hang on to the thing that is *mine*. The research team also found that the aversion to giving up something that is *mine* also triggered activity in brain regions that track feelings of conflict and loss that weren't present when volunteers were considering buying something new. So the brain's value system takes our sense of whether something is "mine" into account when making decisions, even when there's very little practical value to the object's attachment to us.

If the effect were limited to small things that clutter our kitchen cabinets, it might not be such a big deal. But we give up other, bigger opportunities by conflating self and value this way and by being afraid to surrender other things our brains see as ours, like ideas and habits.

## OUR CORE SELVES

We've seen that one reason it can be so hard to part with things is that, to the brain, they can feel like part of our identity and therefore feel "good." The same goes for ideas and behaviors. If we want to feel less defensive and more open-minded, then one way is to recognize which traits and values really are important to our sense of self and which aren't.

Think back to when we first talked about the entanglement of self-relevance and value in Chapter 2. I invited you to think about what you'd say if I asked you to describe yourself. Maybe you said you were thoughtful, ambitious, and punctual. If I now asked you to put those qualities in order of importance—in other words, to rank them according to how "core" they feel to your identity—it might take some moments' reflection, but you could probably do so. For example, I might say that I care a lot about the people I'm close to, and my sentimentality and my desire to hold on to mementos like my mug from Newport are more peripheral extensions of this core trait. Back in Chapter 2, Jenny Slate's desire for connection was core to her personality, and her sweetness extended from that core trait.

In this sense, the way we think about our traits can be thought of as a network, with our most important, core traits at the center. Research led by Jacob Elder and Brent Hughes, psychologists at UC Riverside, mapped perceived relationships between 148 different positive traits in experiments conducted with hundreds of people. To explain the relationship between different traits, the researchers drew a helpful analogy with how we might think of a robin: the feature "wings" might be considered central, since other features important to the robin's bird-ness, like "flying" and "nest building," depend on its having wings, whereas the feature "red breast" would be considered less central, since those other characteristics don't depend on it. In terms of our own human traits, people tend to believe that being "witty" depends on being "fun," that being "fun" depends on being "sociable," and that being "sociable" depends on being "outgoing." The network of these traits, accordingly, would look like this:

Outgoing (core) ➤ Sociable ➤ Fun ➤ Witty (peripheral)

According to this model, on average, someone who thinks of themself as "witty" would likely also say that they are "fun," but someone who describes themself as "fun" might not necessarily say that they are "witty," since fun is closer to the core. This type of networked struc-ture creates a sense of coherence and organization in our self-concept.

Intriguingly, Jacob and Brent's research suggests that the brain takes this networked structure into account when it processes self-relevance. As we know, our self-relevance system allows us to answer basic ques-tions like "Am I polite?," "Am I messy?," "Am I honest?," and so on. But the self-relevance system doesn't treat all these judgments the same way; some come more easily than others. The research team looked at the relationship between brain activation in key parts of the self-relevance system when people judged how much different adjectives described them, and related the resulting brain activity to how much people typically think different traits depend on one another (that is, how central each trait is). Would there be a relationship between brain activity and the centrality of the trait?

There was. Consistent with past research, people tended to say that positive traits that are more core described them more strongly and that negative traits that are core described them least strongly. People who were high in self-esteem tended to do this the most, and people who were more depressed tended to do this the least. The team also found that certain regions in our self-relevance system, like the medial prefrontal cortex, appear to track how deeply ingrained these different parts of ourselves are to us. Thinking about "core" traits recruited less activation of the medial prefrontal cortex than thinking about more "peripheral" ones. In other words, it seems that the self-relevance system might require more processing to make judgments about traits that are more peripheral than to make judgments about the core traits that define us. Just as we have an intuitive sense that our various traits don't contribute equally to our identities, our brains don't see them all the same way either.

Importantly, it doesn't necessarily defend them all the same way either. Jacob and Brent's team also found in another study that it can generally be easier for us to incorporate feedback about traits that are more peripheral than core. In the study, the researchers asked undergraduate volunteers to rate themselves on the 148 different traits they had previously mapped—from "persistent" and "thoughtful" to "prideful" and "inquisitive"—then videotaped interviews in which the volunteers talked about themselves, their goals, and their interests. The study participants were told that their videos would be shown to their university's admissions committee, who would provide some feedback about their personal qualities. In a later session, after being told how the committee ostensibly saw them, they completed the trait-rating exercise again. (In reality, the feedback was randomly generated, so the researchers could separate the way that people incorporate feedback from their actual personal qualities.)

Jacob and Brent's team found that after receiving feedback, people were more likely to change how they rated themselves on traits like "well-spoken" that were mapped as less core in the general semantic network, but held fast to more core traits like "friendly." Unsurprisingly, people were more likely to adjust their self-views to be more positive (for example, *maybe some of my ideas are more original than I realized*) while resisting feedback they perceived as negative (*maybe the committee couldn't tell that I'm friendly because I was nervous*).

As we saw earlier, this bias toward having optimistically positive self-concepts often helps people maintain healthy levels of self-esteem, and the tendency to resist changing our self-views on core traits helps maintain a sense of self-coherence. Sure, I might be messy (a more peripheral trait that few others depend on), but I'm also kind (a core trait). If I think of myself as loving, joyful, and sociable, I can use that information to protect my self-esteem (*okay, my desk is a mess, but I'm still a loving friend*). In addition, people who had higher self-esteem had a higher positive learning rate—that is, they were more likely to update their initial views of themselves in response to the positive feedback—whereas those with more depressive symptoms were less likely to update their self-views in response to positive feedback and more likely to internalize negative feedback.

In a related phenomenon, the average person also tends to take more responsibility when good things happen and less responsibility when bad things happen. When Brett seems like he's having a good time during our nightly kitchen chats, for example, I might pat myself on the back for being a good conversationalist. But when he seems vaguely annoyed, eyebrows furrowed, it's easier to imagine he had a bad day than to consider whether my scrolling on my phone is annoying him.

If what we are trying to maximize is how good we personally feel in the moment (or minimize the potential to feel bad in the moment), then giving less attention to our weaknesses might seem like it makes sense. The problem is, when we sweep our weaknesses and potential areas for improvement under our mental rug, we miss out on the chance to do better. Instead, reflecting on our core traits gives us a potential tool that we can use to grow and change.

## AFFIRMING OUR CORE VALUES

Take a minute to think about what matters most to you: maybe it's your relationships with friends and family, your spirituality, your ambition, or your creativity. Now think about a time in the future when you will tap into that: perhaps spending time with loved ones or working on creating something bigger than yourself. Psychologists refer to this technique of reflecting on your values as *values affirmation*.

In neuroimaging experiments, my team has observed how brief values affirmations change the way people's brains respond to advice and make them more open to adopting new ideas and behaviors. Values affirmations help people see or remember that their self-worth doesn't hinge on a single behavior. For example, if I take a minute to reflect on what is core to me, I can see that I'm still a curious and creative person, even if I used my phone while I was hanging out with Brett—and maybe listening to his feedback and changing is even more compatible with who I want to be than ignoring the feedback would be.

This is essentially what we found in the lab, too. In one study, led by Rutgers psychologist Yoona Kang when she was a researcher in my lab, we worked with volunteers who spent a lot of time sitting—at desks, on the couch, in the car. None of the volunteers had specifically expressed an interest in changing this habit (we were interested in studying defensiveness, so we didn't want to recruit folks who already felt motivated to change). When we brought them into the lab, we gave participants a list of values—these might be compassion, friends and family, spirituality, power, wealth, fame, creativity, or independence—and had the volunteers rank the values in terms of the personal importance of the different values in their lives. We then asked some of the volunteers to spend a few minutes thinking about the values that were most important to them, while we asked the others (our control group) to think about things that were low on their ranked lists—things others might value but that weren't personally important to them (for example, someone who isn't particularly spiritual might understand how religion helps others, but might not view it as central to their own well-being). Then, while we scanned their brains, every volunteer saw the same messages coaching them to get more exercise.

Although all the volunteers were reacting to the exact same coaching messages, their brain responses notably varied, depending on whether they had spent time reflecting on values that mattered to them, personally, beforehand. Compared with the control group, the people who had reflected on their core values showed greater activation in their brain's value and self-relevance systems in response to the coaching. In the month following the brain scan, we continued to send affirmation prompts and exercise tips via text message that matched what the volunteers had seen in the lab. In parallel, we measured their behavior using wristwatches that objectively tracked their activity levels, which they had also worn for several weeks before coming into the lab. The people who were prompted via text message to think about their core values briefly each day before the exercise coaching messages increased how much exercise they were getting by about 5 minutes a day on average, whereas the control group showed a 3-minute average decrease in their daily exercise. This may seem like a small change, but considering that the CDC recommends getting 150 minutes of exercise a week (a little more than 20 minutes a day), this change could be meaningful to their health. In other words, when people spent a few minutes reflecting on the values at the core of their identities, their brains were more open to valuable advice, and they became more likely to change in response.

In other research, people who underwent values affirmation were not only more likely to be open to evidence that challenged their views, but also more likely to consider the views of someone with a different ideology. Smokers who reflected on their core values felt more connected to other people and were less skeptical about the validity of information on the harm of smoking, and affirmed drinkers were more receptive to information about links between cancer and alcohol. Values affirmation exercises have also helped white Americans feel more open to thinking about the role of white privilege in society and to see the role of structural racism more clearly. White people sometimes think of racism as something that other people do, and they might feel defensive when thinking about the invisible ways that racial bias benefits them. After a values affirmation exercise, though, white Americans were more willing to conceive of racism in institutional, rather than individual, terms, recognizing that racial oppression is embedded into many of society's structures as much as it is defined by individuals who are themselves explicitly racist. This way of thinking helps highlight the ways that structural racism creates advantages for white people and disadvantages for others and highlights potential paths to reducing societal inequality.

More broadly, if reflecting on what actually matters to us most can help make us more open to new perspectives and feedback that gets us where we want to go, this makes it a potentially useful tool. It's also a tool I use in my own life. I often try to protect some time before meetings when I think there might be feedback that is particularly hard (but useful) to hear (for example, performance review time when my team and I give each other bidirectional feedback), or when I know I'm going to interact with someone who I anticipate will be critical. I'll use this time to write in my journal or go for a walk and reflect on what really matters to me—or just spend a few minutes looking at the photos on my wall that remind me of things I care about like fun times with my kids or discussions with my grandma. Being grounded in these things helps me in the meeting: I'm better able to hear what others are noticing and separate that from the initial feelings of defensiveness that could cloud my ability to decide if I want to make related changes in response.

If the idea of reflecting on your core values doesn't appeal to you—maybe it feels hard to figure out what is most important to you right now, or you don't like writing in a journal, or you're trying to help someone else who doesn't feel like doing it—there are other options. To create similar effects, researchers have used quizzes where people get to answer questions about parts of themselves they like or choose an identity they think defines them (for example, "artist, comedian/funny person, athlete, musician, entrepreneur, student, nurse, doctor, lawyer, mathematician, scientist, and engineer") and then are asked to insert the term into sentence stems and complete each sentence (for example: "Being a ___ makes me feel ___"; "Being a ___ reflects my true ___"; "When I am being a ___ I experience ___").

In a range of settings and with a range of ways to engage, values affirmation can reduce our instinctive defensiveness and make us more open to change, and as we successfully change, that growth can feel rewarding. So how can we make sure these values affirmations have the biggest impact?

Remember when I said that I sometimes keep some time for myself *before* a potentially difficult meeting? There's a reason for that: Values affirmation exercises seem to work when they are delivered right before potential threats. They can also work immediately afterward, but the window of effectiveness is short—sometimes only moments. Once we've had a chance for defensiveness to start to crystallize, it's often too late for affirmation to work. For example, if you're in a meeting and find out that someone doesn't like your idea, and then someone else begins to ask you questions about that idea, you often don't have the opportunity to effectively use values affirmation—defensiveness has likely already kicked in. This means that if you're a doctor delivering advice to a patient about changing behaviors related to their health, it could be helpful to give your patient a chance to reflect on core values that matter to them right before you talk to them or right after you give them suggestions for change, but probably not once you've asked them to give you a response to the advice. Psychologists Geoffrey Cohen and David Sherman describe social psychological interventions like values affirmation as being "like an engineered coincidence. It places in close proximity three events that otherwise might seldom co-occur: a positive influence, a challenge, and an immediate chance to change."

Another piece of good news (or bad, depending on whether you are in a position to intervene early) is that small wins (and losses) can snowball. A meta-analysis of fifty-eight studies found that values affirmation improved the academic potential of students from marginalized groups in a variety of ways, ranging from the benefits of values affirmation before a big test to its effect on grades in school. Researchers at Columbia University, the University of Colorado, and Stanford saw this in one study that followed Black students who received values affirmation interventions in the first week of school; they subsequently experienced an improved sense of belonging and maintained their academic momentum. However, the students in the control group, who did not have the early values affirmation intervention, experienced a dip in their sense of academic belonging, as well as in their performance. After these students later received the values affirmation intervention, their grades sometimes improved, but they did not recover a sense of belonging, even if they performed well. So *when* these interventions happen, in relation to a challenge, can impact their effect—and that can have important consequences.

We can use this information about the timing of these interventions to consider when it might be helpful for us to do our own values affirmation or provide the opportunities for others. Times of transition—for example, the start of high school, the start of college, the start of a new job—are moments when things feel less predictable and people tend to feel greater threat. Even brief values affirmation interventions at times like these can have outsized effects, since early wins can build confidence and lead to advantages that accumulate. By the same token, if we don't pay attention to what happens early on (when onboarding a new person to our team; as our kids start at a new school), negative feelings can also spiral, and the effects can be difficult to later overcome. These temporal effects are part of a bigger picture where the effects of personal engagement with affirmation also interact with the environments we find ourselves in. An individual affirmation exercise won't undo the damage of a toxic work or school environment, but if delivered early on, it can complement other efforts that boost a person's sense that they belong and are valued. Early success can set people up for later, and repeated, success, whereas early negative feedback can do the opposite.

Repeating these values affirmations can also create a positive feedback loop: doing them regularly can contribute to building a larger sense of purpose, which can also reduce how threatening we find feedback suggesting we change. My friend and colleague Vic Strecher, an expert on the science of purpose, defines it as "goal-directed behavior around what matters most," which "becomes a central, self-organizing, life aim." Studies show that a strong sense of purpose has many benefits: people who report having a strong sense of purpose tend to live longer, happier, healthier lives, suffering less frequently from heart disease and cognitive impairment and experiencing fewer strokes than people who report feeling less purposeful. Scientists believe this correlation between purposefulness and longevity may in part be due to a tendency among people with a strong sense of purpose to take care of themselves. In turn, research shows that engaging in healthy behaviors—like calling a good friend, getting a good night's sleep, or going for a walk—can help boost purpose.

Inspired by this work, the team in my lab working on values affirmation and exercise coaching also wondered whether feeling a stronger sense of meaning and purpose in life might make it easier for people to hear coaching and other constructive advice as helpful. In this research, again led by Yoona Kang, we first asked participants to rate how strongly they felt they had a larger purpose in life before having them enter the brain scanner and receive health coaching. Yoona discovered that people who reported feeling a stronger sense of purpose indeed showed less activation in brain regions that detect conflicts when they were exposed to the coaching messages. It's not entirely clear yet why this is: it could be that a sense of purpose makes it easier to see how short-term actions can support longer-term well-being or other goals. Alternatively, feeling a sense of purpose might be grounding, giving people the confidence to feel less threatened and defensive when asked to change smaller things about their behaviors.

If we want to maximize the effects of values affirmation or other exercises that help connect us to our bigger goals or purpose, growing evidence suggests that values and purpose that connect us to other people are particularly powerful. When my team studied the effects of values affirmation on the brain, we let participants in our experiments choose values from a long list—they chose the ones that were most important to them. As we've seen, connecting with these values—whatever they were for each person—helped the volunteers in this group receive the exercise coaching as valuable and relevant to them and subsequently change their behavior. But we wondered: might some values have worked more effectively than others to break down defensiveness?

When we looked at what values people had ranked highly in the values affirmation exercise, and what their brain activity in response to the coaching looked like, an interesting pattern emerged. Compared with people who chose more self-focused values, like fame, people who had chosen what we call "self-transcendent" values, like friends and family and compassion, showed lower activation in brain regions that track threat, like the amygdala, as well as a broader set of brain regions that were activated in prior studies that manipulated people's sense of threat in the scanner (for example, looking at scary images). Identifying the reasons why self-transcendent values are particularly potent in opening us to change is an active area of research, but one possibility is that in shifting our focus to our connections with others and to their well-being, we let go of some of the default narrow self-focus (and narrow self-definition) that produces defensiveness in the first place.

## WITH A LITTLE HELP FROM MY FRIENDS

On the night Brett brought up the cell phone issue, I did feel defensive at first. It wasn't until the next day that I started thinking about other people in my life and how they interacted with their phones and with me. I was able to think of the friends who never use their phones when we hang out and how connected I feel when I'm with these people. I thought of delicious dinners in grad school with my friend Erin; deep and penetrating conversations with Alex and Clara in their garden and art studio; of long, rambling walks with Niaz. Absent from all these interactions was a little rectangular interruption machine. They were just there—present—with me. I knew I wanted to be like that for my partner.

For many of us, being connected to other people is one of the most important and meaningful things in our lives. As we explored in the first part of this book, our understanding of what others think and feel shapes us. Beyond opening us up to new ideas and behaviors, seeing what other people value and how they behave can change what we ourselves value and how we behave. Even *stories* of other people can play a powerful role in changing our thoughts and actions. Stories tap into the brain's social relevance system. As with all paths to change, the changes spurred by these kinds of social learning can make people happier, healthier, and more connected—but they can also lead to outcomes we might not choose for ourselves, behavior that leads to harm and even violence. By understanding how they work, we can be more aware of what's unfolding, and why.

Learning from role models, whether in real life or in stories, can bypass the defensiveness that often arises when we are confronted with more direct persuasive messages or attempts to change our behavior. At work, for example, I often think to myself, "What would Karen Hsu do?" Karen was my boss at the healthcare consulting company where I worked after college, before I got into cognitive neuroscience. Karen was a great boss—she taught us the value of a silent wink and a smile to our team when she wanted backup on a new idea she was about to pitch on a group conference call with all of us huddled around the phone in her office. She always treated us at happy hour and was generous with her time, showing genuine interest in getting to know what we valued, what we were good at, and where we needed to grow. She rarely asked us to do things that she hadn't done herself, and she was great at helping chart realistic paths to success. Importantly, she was open-minded and took our feedback seriously, even though everyone on her team was just out of college. When, as a lab director now, I find myself sitting across a table from a student in tears over the stress of a failed exam, or a flustered staff member who just had a difficult interaction with a vendor, or a colleague who has an idea about how the lab could or should do things better, I think of Karen. It helps me stay calm, warm, and focused on what is good for the individual and the team.

Research by psychologists Meghan Meyer and Diana Tamir suggests that when I think about Karen this way, it might actually change the way I see myself. In one of their studies, Meghan and Diana asked volunteers to recall things that had happened to them in the past—for example, getting good news or doing poorly in school—and how they had felt, then to imagine how a friend would feel in the same situation. When the researchers later prompted the volunteers to think again about how they themselves had felt, they saw that the volunteers' ratings of their emotions in the situation had changed: now they moved toward how they had imagined their friend would feel. Likewise, in another study, when volunteers were asked to rate their own personality traits after thinking about a friend's personality, their answers changed from how they had rated themselves earlier, growing closer to their friend's traits. These effects were strongest when the person the volunteers were thinking about was someone they already perceived as like themselves. In this way, imagining how Karen might approach mentoring and receive feedback might really make me more generous and open-minded like her.

Neuroimaging research has further illuminated how, when we are taking the perspective of a person with particular traits, our brains respond as if we had those traits as well. Michael Gilead and Kevin Ochsner ran a study at Columbia University in which volunteers learned about two "peers" who had ostensibly participated in the study before them. In reality, these "peers" were made up by the study team to evoke a stereotypically tough person and a stereotypically sensitive person. The "tough" peer worked as an emergency medical technician, loved horror movies and loud music, and described themself as resilient, while the "sensitive" peer worked as a graphic designer, loved romantic comedies and classical music, and described themself as sensitive. After learning about the two peers, the real volunteers were asked to rate different pictures from the perspective of each peer, as well as their own perspective, while their brains were scanned. Some of the pictures were neutral (think images of furniture and everyday objects), and some were chosen to evoke strong negative emotions (think scenes of blood and gore). As you would expect, the negative images were rated as more negative by volunteers than the neutral images—but *how* negative the volunteers rated them depended on whose perspective they were taking. When they took the perspective of the sensitive peer—predicting how that person felt while seeing the image—they rated the images as more negative than when they took the perspective of the tough peer. Taking the perspective of a sensitive peer also made the brain's emotional response more sensitive, whereas taking the perspective of a tough peer made it less sensitive.

Within the volunteers' brains, the self-relevance and social relevance systems also tracked whose perspective they were taking. Using patterns of activation within the part of the medial prefrontal cortex at the intersection of the self-relevance and social relevance systems, the research team could differentiate and predict whose perspective a person was taking at a given time. In addition, activation in the social relevance system correlated with activation in regions responsible for emotional responses—suggesting that taking the perspective of a sensitive or tough peer activated the social relevance system and also changed the person's own emotional responses.

In this way, adopting someone else's perspective can help us approach situations in our lives that are difficult or may be threatening to our sense of self. Much the way I ask myself "What would

Karen Hsu do?" when I'm confronted with a tough work situation, we can try to think of events in our life from the perspective of a friend, mentor, family member, or character in a story who would react with the values we aim to embody (does this situation call for responding like *Star Trek*'s Mr. Spock or Captain Kirk?).

This can apply to other domains of life as well. Receiving news of a job rejection, having our doctor tell us we need to adopt a healthier lifestyle, or getting feedback from a colleague on how our tone may have come across in a meeting—all of these might feel less threatening if we try to see it from the perspective of someone who isn't us, but who would react in a way we may want to. For example, a friend who was trying to follow her doctor's advice and eat more vegetables recently called me—someone likely to react to a veggie-heavy diet with excitement. She was right, and I was happy to share my favorite easy and fast veggie dinners—pasta with roasted eggplant, red sauce, spinach, and goat cheese; hearty Caesar salad with avocado, toaster-oven croutons, and apples; veggie coconut curry and rice. As a bonus, we got to spend some time together cooking and eating. In this way, taking her doctor's advice provided a chance for us to connect, which I hope offset some of the hard parts of trying something new.

Beyond changing how we feel, taking someone else's perspective can also change what we do. Once you've taken in feedback, perspective taking that taps into the social relevance system can continue to be a useful tool for changing behavior. In a randomized experiment that was part of a large, interdisciplinary collaboration that I led involving several labs at the University of Pennsylvania, Columbia, Dartmouth, and Brigham Young University, college students drank less alcohol when they were reminded to take the perspective of a peer who drank less than they did. Volunteers in twenty-four social groups at the University of Pennsylvania and Columbia University rated one another along a number of dimensions, including how much different people drank alcohol. For the month that followed, we sent the students text messages every day coaching them about how they should approach alcohol that day. During some weeks, they were told to react naturally to alcohol if they encountered it that day. During others, they were told to take the perspective of a peer who drank less than they did; for instance, if Mary reported during the social network survey that John drank less than she did, we would later suggest in the daily text message reminders that Mary approach alcohol the way that John would.

During weeks when the students were reminded to take the perspective of a peer who didn't drink much, they reported drinking less than the weeks when they reacted naturally. In a first study, while students were on campus, the intervention reduced the reported frequency of their drinking from about one night in six to one night in eight. In a second study, when students were off campus at the start of the COVID-19 pandemic, the intervention reduced the reported frequency of drinking from about one night in nine to one night in eleven. This means that a simple intervention that focused students' attention briefly on a peer who drank less than they did each day changed their reported behavior in meaningful ways, and all it took was a small text notification. The people we pay attention to shape who we become.

### TAKING A STEP BACK

It's not just the social relevance system that responds to values affirmation and perspective taking. We can also tap into the self-relevance system with these approaches, opening ourselves up to change by increasing how self-relevant and valuable we find new ideas and behaviors. But if, as we've seen, defensiveness arises from holding on to our core self too tightly, another approach suggests itself: removing our self from the situation.

Many of us have an intuitive sense of how helpful taking a third-person perspective can be. The psychologist Ethan Kross often illustrates this with an example from basketball history, when LeBron

James had to make a difficult decision about whether or not to stay with his hometown team, the Cleveland Cavaliers. LeBron had become a superstar with this team, and Ohio fans loved him, but he hadn't won a championship. Now he had the opportunity to join the Miami Heat—a team with excellent players and likelier championship prospects. When talking about his decision on television, LeBron explained: "One thing I didn't want to do was make an emotional decision. I wanted to do what's best for LeBron James and to do what makes LeBron James happy."

Ethan highlights that LeBron started out referring to himself in the first person ("I"), but switched to a third-person perspective ("what's best for LeBron James"). Ethan's research shows that talking to ourselves in the third person is an effective way to regulate negative emotions. ("Take a deep breath, Emily. That's a lot to handle, but you can do it.") His research also shows that describing problems in the third person (in other words, creating some distance between "me" and the person who is experiencing the problem) can increase our ability to solve problems. Even taking the perspective of a "fly on the wall" helps us feel less angry and reduces aggressive thoughts and reactions that would otherwise escalate the conflict. Indeed, these techniques are forms of reappraisal, which we saw in Chapter 4 can be used to either increase or decrease negative or positive feelings.

Thinking about situations from a new angle or perspective and letting go of our tendency to interpret situations in terms of their personal relevance to us actually changes how the brain works. In the same large study in which we taught college students to take the perspective of a peer to reduce their drinking, my team also explored what happened in people's brains when they were coached to react to pictures of alcohol (and later to real alcohol in their daily lives) a little differently. We asked them to take a step back to "mentally distance themselves by observing the situation and their response to it with a more impartial, nonjudgmental, or curious mindset, and without getting caught up in the situation or response." We then compared what happened in the brain when volunteers responded this way with what happened when they reacted as they naturally would.

Although it took effort for the students to control their brain states and think impartially, once they did, communication between regions across their brains unfolded and changed more quickly. That is, when people took a step back and stopped judging all their thoughts and feelings, they were literally less stuck on whatever brain state had just happened and more able to move on to the next.

I thought of this study in a recent yoga class when the teacher urged us to "trust that the moment will move," as in, when you feel discomfort, instead of judging it or freaking out, trust that time will keep moving forward and things will change. *Trust that the moment will move.* I loved it. When I'm feeling stuck, grumpy, or generally not loving how I'm feeling, I like to think about this study and about how just letting go might help my brain be more dynamic.

Indeed, as we touched on in Chapter 2, the idea of taking a step back and approaching things with a nonreactive attitude—what scientists call "psychological distance interventions"—also aligns with thousands of years of wisdom in Eastern traditions, including not only yoga but other forms of meditation and mindfulness. These traditions, and complementary modern research, have long emphasized that by letting go of a bounded notion of the self, of what is "me" and "not me," and embracing the variability inherent in our day-to-day lives and selves, we can become less reactive to what we might perceive as threats to the self. Research on people who meditate seriously, over long periods of time, also suggests that it is possible to change our brains' automatic responses to everyday events. Even briefer experiences that teach people how to engage in mindful acceptance or shift the focus of their thoughts have been shown to reduce physical pain in response to touching a painfully hot object, and negative emotions in response to distressing images (as reflected by both self-reports and brain signals), as well as cravings for alcohol, cigarettes, unhealthy foods, and other drugs. Mindful acceptance can help us regulate emotional intensity by changing our appraisals of the emotional significance of different images and experiences. If we let go of a strict notion of the self, then there is nothing to defend.

Our sense of "self" is built in part from our past experiences, beliefs, and behaviors, and we often cling particularly tightly to these fragments in an effort to maintain a coherent and positive sense of who we are. But if we want to make choices that are best aligned with what we value, sometimes it helps to let go of the fragments of "self" that aren't serving our current needs and future goals.

I missed chances to connect more deeply with Brett when I failed to respond to his earlier suggestions that I stay off my phone in the evening. Similarly, we could lose out on the chance to learn new skills and become better at our jobs when we dismiss critical feedback from colleagues or team members, or lose out on the chance to connect with our family and friends when we justify the annoying thing we did or blame the other person, rather than trying to see how we might have been responsible for a conflict.

Observing a situation and responding with a nonjudgmental, curious mindset can take the emotional edge off the prospect of change and literally help our brains be more nimble and present-focused, rather than stuck on defending past ideas of who we are. We can get that sense of distance by taking a third-person perspective or taking a step back ourselves, which can help reduce the impact of defensiveness and emotions that cloud our ability to change. But rather than just becoming more open to new ideas, opportunities, or changes as they happen to cross our paths, can we go a step further? How might we imagine new ways of actively connecting with a broader range of people and ideas?

# 6

# Connect the Dots

IT WAS A SWELTERING late spring day in Tonya Mosley's middle school classroom. There was no air conditioning in her Detroit public school, just fans blowing hot, sticky air from one place to another. Many of the windows were sealed shut. The kids were sweaty and uncomfortable—not ideal conditions for learning. Yet the school's dress code required girls to wear shorts that extended below their knees. *Why?*

The question stuck with Tonya all day, and as part of her daily report to her mom when she got home, which generally included news of what was happening in the classroom, Tonya mentioned the unfair dress code.

"That's just how it is," her mom said. "It's always been this way." But that didn't seem like a good enough answer to Tonya.

Tonya knew that she occupied a unique position at school. She was well liked by her classmates, her teachers, and the principal, who all thought of her as kind and honest, and she loved talking with people and using the information she observed and that they shared with her to help others. She had a sense that if you could just get people to understand one another's perspectives, you could make real change.

Moved by this feeling, Tonya wrote a petition to update the dress code and got nearly all her classmates to sign it. Then, Tonya went to school wearing short shorts that she knew would be noticed. When they were, she gave the petition to the principal and explained the students' perspective—that the dress code was exacerbating the bigger air-conditioning problem, and its effects were most unfair to girls.

What happened next opens a window into the potential and limitations of the social relevance system. We've already seen how others can influence our decision-making—from choices as mundane as whom we regard as attractive to choices as consequential for the planet as how effectively we conserve energy. We've also seen how strong an input social relevance is into the value calculation. Indeed, scientists believe that one of the reasons humans evolved such large brains relative to our bodies was to help keep track of complex social information; the fact that people live in social groups, which helps us survive, means that we need to coordinate with others. But we also tend to limit ourselves to people who are most like us and who validate our existing beliefs and preferences. Adults, for example, might listen more readily to the opinions of other adults and might not even be aware of the concerns that kids have.

This may have been the case for Tonya's principal. He hadn't been paying attention to the students' perspective on the connection between the dress code and the air-conditioning problem. Then Tonya arrived in his office armed with a petition and a new point of view. By helping him see things in a different way and using the relationships she had built, Tonya was able to catalyze a change. Her principal updated the dress code.

Later in life, Tonya became a journalist—if you regularly listen to National Public Radio, you have likely heard her. She has served as the Silicon Valley bureau chief for one of America's largest public radio stations, San Francisco's KQED, and then as the host of *Here and Now*, a news show in Boston that reaches upward of five million people each week. You might have heard her interviewing guests on

*Fresh Air*, which she cohosts with Terry Gross, or maybe you're a fan of her podcast *Truth Be Told* or the incredibly powerful and personal series *She Has a Name*. Some of the skills she uses most in this work are a natural extension of the ones she used to change the dress code: she connects readily with a wide range of people, listens to what is important to them, and helps translate that viewpoint to audiences that might not otherwise have access to that perspective.

Sociologists call people like Tonya "information brokers"—people who connect with a range of others who don't otherwise interact. In business organizations, information brokers tend to get paid more, and promoted faster, in part because they have access to more diverse sources of information and ideas that they can leverage to solve problems. They also tend to be seen more readily as leaders.*

Although the relationship goes both ways—business success brings new opportunities for connection—we can also make change by building bridges between different people and groups who might not already understand one another's perspectives. In these efforts, the social relevance system, which helps keep track of people and the connections between them, is key. In this chapter, we'll explore how the social relevance system automatically keeps track of people's roles and status and how we might broaden the beam of our flashlights to include more people's perspectives when we make choices and build connections. This is crucial because the people whom we consider relevant, and whom our social relevance and value systems pay attention to, shape what ideas get priority in our minds. But they also *limit* what ideas get priority. We might see and have access to more oppor-

---

* Notably, these advantages are greater for people whose identities fit societal stereotypes about the kinds of people who are leaders. In one recent study, the advantages for men who were information brokers were twice as large as for women in similar positions. The exception was women who achieved their brokerage status through communal efforts, like keeping in touch with old connections while building new ones when they transferred from one unit to another. One theory is that this is because that behavior is consistent with societal pressures for women to behave in communal, caring ways.

tunities, for ourselves and others, if we expand the way we think about whose perspectives are most relevant.

There are many ways to accomplish this, and in this chapter we'll meet people who each go about this quite differently: Tonya, who does this work by highlighting a wide range of voices in her journalism; Roland Seydel, an "innovation explorer" at Adidas with whom I worked to build successful intracompany communication; and Dani Bassett, a fellow neuroscientist who has studied (among other things) curiosity. These elements—listening to wide-ranging expertise; communicating across different spheres; and being curious—all come together to potentially transform the ways we connect and find value in connection.

## WHO KNOWS WHOM?

When I met him, Roland's job was to come up with new ideas and help teams design products that would change the field for a wide range of sports. He was trained in physics and sports education, as well as mechanical engineering, and he wanted to make it easier, more comfortable, and more enjoyable for everyone to play sports, to move. The culture at Adidas was focused on engineering excellence—they made high-quality products that often looked great and worked superbly. But they also could be seen as having a limitation: products sometimes seemed to lack a cohesive message that obviously connected their excellence with what the consumer wanted. For instance, the engineers might have worked hard to make a new soccer cleat lighter (thus helping the wearer run faster), but if the average teenager saw the shoe in the store, that wouldn't necessarily be obvious to them. This meant that on a practical level, I wondered if products were always getting into the hands of the people who would most benefit from them.

Roland had realized that an important part of Adidas's issue was that people in different areas and departments might not know what the others were doing and didn't always speak the same metaphorical language. This was the question Roland was trying to answer when he brought me on board as a consultant. Together, we needed to develop a kind of Rosetta Stone that would allow people across the organization to focus on what was most important—and, most critically, to talk to one another about it.

In describing his approach, Roland told me that he wanted to look at problems through the "eyes of an insect." Their compound eyes featured thousands of tiny, independent facets. He believed that all these angles, all these perspectives, could come together and form a more complex whole than if we looked at the world through one, simple lens. So he, like Tonya, worked to gather up a wide range of perspectives. He was the perfect person to help people across Adidas to talk to one another about product design, innovation, and marketing from the perspective of their field of expertise.

Indeed, connecting people with different kinds of ideas and expertise is key to creating the benefits that Roland was after. Research shows that being able to pick up the phone and call people who have different kinds of expertise and perspectives helps us solve problems creatively and productively. To do this well, Roland had to simultaneously know many different pieces of information himself, understand what others knew, and recognize where they might have gaps. He had to build a mental map of who knew what, and who was connected to whom at Adidas. He needed to know whose expertise was complementary, who might already be in conversation, and who might need a more explicit introduction to get a conversation going.

You might imagine that Roland's background in physics could help him here: the kind of smarts that enables him to map how objects relate to one another might also translate to effectively visualizing graphs and noticing holes in the social fabric, sussing out fruitful opportunities to connect others. Or you might suspect that other skills are needed to mentally map and interpret the social relationships between people in our orbits.

Brain imaging can help us distinguish between these different possibilities. Back at Penn, I worked with a team led by Steve Tompson and Dani Bassett to understand which brain systems allow people to learn the structure of a social network. The team brought college students into the lab and scanned their brains while they saw sequences of avatars: the avatars represented people, and their sequence represented their relationships. By observing these sequences, the research participants learned the structures of social networks they had never seen (that is, relationships between people they didn't know)—a simplified version of the way Roland might learn about the relationships at Adidas. The students also learned the relationships between objects. We found that some of the same basic memory systems that help people learn and remember relationships between different objects are also used for understanding the relationships between networks of people.

But how the brain tracks these relationships isn't identical. Notably, learning the structure of who is connected to whom—which *people* go together in a social network—involves the social relevance system in a way that learning about the relationship between nonsocial objects like, say, different types of building materials or rock formations doesn't. Being able to reason about people and being able to reason about things are complementary but distinct skills. Some people are great at learning nonsocial network structures but lousy at learning who is connected in a social network, and vice versa. If we want to develop our ability to connect different people and ideas for the sake of innovation, we may need to do so using different brain systems than we do when we focus on relationships between objects.

One starting point for doing this is to more proactively notice whom your networks include beyond the people who immediately come to mind. In many businesses, people's relationships tend to fall into predictable, tight-knit clusters according to their roles and business units. When we move to a new unit or role, making an effort to keep in touch with people from our prior job is one path to building useful bridges. Likewise, the people who take active charge of communicating information between people who aren't already connected to create shared understanding tend to be seen as leaders.

In one study, executives who went through a training program aimed at teaching them how to map the connections in their organization and how to take advantage of these connections were promoted more often and rated higher in performance evaluations—but only if they were active participants in the training. By more actively taking time to notice or map out who is connected in your networks, you can also practice identifying who might *not* already know one another and get better at connecting them—or connecting ideas from these different sources, even if the people themselves never interact. These skills play out in business organizations, online communities, and research communities. Similar skills might also be valuable to artists developing their creative work, government officials or citizen assemblies coming up with new ways to solve problems in their communities, or school administrators thinking about new programs.

After considering the psychology and neuroscience literature on motivation and behavior change, and after many iterations, Roland and I came up with what we called the RISE framework. It highlighted how Rational, Identity, Social, and Emotional factors influence people's decision-making. We suggested to the engineers that when designing a product, it was important to understand not only the physics of how a ball makes contact with the toe of a soccer boot (the R in RISE) but also why a person might want to play in the first place. How might we design products that would support their identity and their social and emotional needs as well? Conversely, for folks in marketing, we emphasized the importance of understanding the ways that the carefully engineered products were actually better.

Roland began sharing RISE with people in areas from football (soccer) innovation to women's running to consumer insights. Philip Hambach, who was the director of global consumer insights, remembers how excited he felt seeing the heads of Market and Global together at a whiteboard: "You could see people we wanted to talk to each other, sitting looking at a whiteboard together, debating." These were people who Philip perceived had not previously always found common language but now had a way to learn about each other's perspectives, as well as thinking about their consumers in more holistic ways. They were deep in conversation, sharing insights. With Philip and Roland's support, RISE took hold in more areas of the company and was spotlighted at the Adidas global marketing meeting in 2016.

With this framework, we had found a way to bring people together and provide a common language to discuss their perspectives. But if someone like Roland isn't playing matchmaker, how do we choose where to invest energy? How do our brains sort through the many people around us? Given the large number of all the possible people and relationships that we might want to keep track of for various reasons, how do our brains prioritize which ones to pay attention to? And where do these priorities come from?

## WHO HAS STATUS?

Tonya Mosley has found herself curious about related questions throughout her career. Her work making her podcast *Truth Be Told* was in part a response to her years of working as a television news reporter—difficult, stressful work to which she had been drawn for a long time but that also felt incomplete. After much success in the media industry, she earned a fellowship at Stanford to work with social scientists investigating the implications of implicit bias in journalism. That's when it began to crystallize for her. "The truth is, there is a narrative that is replicated over and over again," she told me—but, she added, "there are other parts of the story that we're missing."

The stories we hear in the media and that we tell ourselves are often scripted according to certain cultural norms, values, and assumptions. The voices that are elevated are determined by certain rules, written and unwritten, about whose perspective on a story matters—rules developed in large part by straight, cis, white men. For example, a news story about the drug crisis in 1980s Detroit might call on economists to comment on the cost of the drug war and the government response, and the resulting story might ignore the actual lives that were lost or the way the drug issues affect the community. Stories like these did not fully reflect Tonya's perspective as a Black woman (nor a bevy of other perspectives).

Although most of us aren't bound by the rules of journalism, parallel judgments come up in our daily decision-making. We saw earlier how strongly the social relevance and value systems are influenced by other people's beliefs, preferences, and actions. My views of Benedict Cumberbatch's attractiveness were partly shaped by the views of people around me. Importantly, this same logic also extends to many other kinds of assessments of other people. Beyond whether we judge them to be physically attractive, are they likely to be trustworthy? Kind? Good leaders? Our brains track these different kinds of social status, and research suggests that we may be more sensitive to the perspectives of people who are valued by others in our networks. In other words, we don't pay attention to people, or the connections between people, equally; instead, we prioritize understanding what some people in our networks think and feel more than others.

Suppose I asked you to think about your experience in elementary or high school, for example. You might still be able to tell me which kids were especially kind or widely liked, even if you couldn't draw a complete map of all the kids in your grade or recall exactly who was friends with whom. If you think about it, the same might also be true for people you work with now (for instance, you may be more aware of the opinions of people with particularly high status in your organization). Even though most of us don't consciously keep track of the exact configuration of our social networks on a day-to-day basis, some of this information—about how the people in our lives relate to one another and to us—is implicitly encoded by our brains. In addition, social media algorithms connect us to some people and not others, and highlight some messages and not others. This makes it even more important to understand this tendency of the brain—so we can become more aware of the ways that the information we're exposed to is likely biased.

The way our brains don't treat all social relationships equally can have significant implications. Imagine that Brad is trying to put together a team at work, and he knows he'll need a data analyst. He scrolls through the company website, considering the possible folks who could do the job. As he scans the faces of the analysts on his team and others, he thinks about the projects they've worked on recently and how things went. When he gets to Jake, he nods his head. Yes, Jake would be great. How does Brad know? Although Brad might not think about all the factors that go into his decision consciously, under the surface, his brain has been keeping track of lots of information about the people he's scrolling through. Who has a reputation for being conscientious, difficult to work with, smart, collaborative? Who is closest to him in the social network? Who is best connected to others who could help spread the word about the product later?

In addition to tracking who knows whom, our brains also automatically track information about the nature of those relationships, such as how popular and well connected others are or who is most likely to provide social support and empathy to others. In one study led by Stanford psychologists Sylvia Morelli Vitousek (who now works at Instagram) and Jamil Zaki, activation increased in the value and social relevance systems when people looked at the faces of peers who were sources of empathy and social support to many others in their social networks. Similarly, the fact that many people think of Jake as both capable and collaborative shapes Brad's initial reaction to seeing his photo on the website. Other people's opinions shape our own social relevance and value calculations—and we may also more readily make a stronger effort to understand some people's perspectives more than others'. For instance, research shows that we are more likely to think about the thoughts and feelings of people who are like us, whereas we sometimes fail to consider that people who are different from us have valuable perspectives too. This might feel unfair, but there are reasons it makes sense for our brains to factor people's reputations into our social relevance and value calculations. For example, Sylvia is certainly an empathy hub in our network of fellow psychologist friends. She's one of the best people to go to when something good happens and you want a friend to get excited with you, and when something bad happens, she knows just how to commiserate. I watch her interact with other people with a good dose of admiration (trying to figure out how she always knows the right things to say!), and I marvel at what an authentic person she is. This tracks with her findings. If you behave in these ways, other people's brains pick it up.

In other words, these signals sometimes contain useful information—a person who is widely perceived as empathetic might be a good person to make friends with. At the same time, there are a lot of reasons why someone might not come to mind when peers are asked whom they like or go to for empathy and support, and these may have nothing to do with how well they might do as a friend or teammate in these situations. Brad might scroll right past the faces of people whom he doesn't know and who are less well connected, but equally capable. Because of that, those junior analysts don't get the chance to prove themselves on Brad's next project. In this scenario, Brad would also miss out on the chance to get to know someone new, someone who might bring a different perspective to his team.

If we're naturally disposed to pay more attention to the perspectives of some (higher-status or more similar) people in our networks, what might we be missing out on? The perspectives of those who may not come as easily to mind might stimulate our thinking, reveal new opportunities, and shine light on new possibilities. In other words, one way to view the function of the social relevance system is to see it as pointing toward the most relevant people, but the flip side of that is that it risks maintaining and perpetuating bias; this is what Tonya was actively working against. A meta-analysis of over fifty studies, led by neuroscientists at the University of North Carolina, found more brain activation in the social relevance system when people thought about people whom they perceived to be part of their own in-group, compared with when they thought about people whom they perceived to be part of their out-group. This is in conversation with another body of research highlighting that snap judgments about a person's gender, race, and class, among other things, influence our perceptions of their expertise in several domains, including their ability as engineers, their qualifications as expert witnesses in court, their general competence, and their collective team performance. And yet studies also suggest that the ideas published by people from underrepresented groups tend to be more novel and sometimes more generative in their fields. What can we learn if we broaden our imagination about whose perspectives might teach us?

This was a question that Tonya, too, sought to answer. I came to Tonya's work listening to *Truth Be Told* during the COVID-19 pandemic. I was first captivated by her exploration of how to hold on to pleasure and joy when the world feels like a dumpster fire and when we know people around us are in pain. I found that in her reporting, and in her podcast, Tonya offers new and helpful ways of thinking about problems like these, big and small, personal and political.

Questioning the mainstream idea of "experts," like news pundits who offer opinions without explicitly calling attention to their subjectivity, Tonya invites people she calls "wise ones" onto her show. These are people who have both deep knowledge *and* lived experience. Tonya explicitly highlights and names that lived experience on her show and intersperses it with other forms of expertise. In doing so, she expands the listener's ideas of what kinds of knowledge and experience are socially relevant. By actively foregrounding the perspectives of people I hadn't heard highlighted on other programs, she helped me think about things in new ways.

On the first episode of *Truth Be Told*, for instance, Tonya interviewed her grandmother, Ernestine Mosley, about her lived experiences of pleasure and joy (in parallel with confronting racism and constrained resources and opportunity), as well as adrienne maree brown, author of *Pleasure Activism*, a book about the connections between sex, pleasure, and the pursuit of social justice. Tonya edited together parts of these interviews and made links between their perspectives with her commentary. Many of us might not immediately connect conversations with our grandparents with conversations about sex, but Tonya did. The conversation made me think about joy and positive emotion in my own life and work in new ways. Much of what we study in my lab can be emotional (for example, preventing cancer, racism, climate disaster), but listening to this episode made me think about new ways of thinking about, and building in more explicit practices of, care and joy as we do this work. Tonya didn't just know that her grandma and writer adrienne maree brown had complementary expertise in creating joy and pleasure in life and in using that to fuel resilience—she put their ideas in conversation with one another to highlight links for the rest of us.

Tonya's work highlights the perspectives of people of color and those whose expertise has been less recognized in mainstream news reporting. Her approach to connecting with multiple people, identifying their different forms of expertise, and bringing them together to spark new ideas is another example of the well-documented and powerful strategy for encouraging innovation that we began to explore earlier. When people are connected to different social circles or interact with others who have more diverse knowledge and expertise, they can come up with more creative ideas and solutions to problems. We benefit by broadening our curiosity, and society benefits when more voices are included in thinking about how to solve problems. And yet many of us miss out on opportunities to make these kinds of connections when we don't see others as socially relevant or when we get so busy that we stop paying attention to the range of perspectives we might have access to.

As different as their backgrounds and careers are, Tonya and Roland share a reverence for expertise and insight from different sources, as well as a recognition that societal constraints often give more attention to some perspectives than to others. In the face of this problem, my Penn colleague Dani Bassett (who led the study about how people learn social networks) and Dani's twin, Perry Zurn, encourage us to actively audit whose ideas we are paying attention to and building off when we solve problems. They have worked with teams to build tools for academics to estimate how much of the research in a given manuscript is authored by women and by men, and by authors of color and white authors.

You can do this too: Collect some objective data about the sources of ideas you are exposed to most. What patterns do you see if you make a list of who wrote the last ten books you've read or who hosts the podcasts you've listened to in the past few months? How about if you make a list of the people you've talked with most at work in the past six months?* Whose ideas are getting the most priority going into your brain? What opportunities for connecting people and ideas, for making connections, might you notice if you expand your assumptions about who has relevant ideas or expertise for understanding the world and the problems you want to solve in your daily life? The more power you have in a given environment, the more important it is to perform this kind of questioning and evaluation.

## POWER GOES TO YOUR HEAD—AND CHANGES YOUR BRAIN

Tonya now has her own production company, TMI Productions, and her role as a person in power extends beyond the stories she covers and how she covers them. Now, she is also a boss. In this role, it can take

---

* There are many online tools to help you analyze your personal network, like Marissa King's AssessYourNetwork and Adam Kleinbaum's Network Analyzer.

conscious effort to remain aware of the power structures in the teams she manages and to foreground the perspectives of others.

For example, the team that makes her podcast, *Truth Be Told*, is highly communicative, but as the boss, Tonya is mindful of when she sends messages. She knows that people will feel pressure to get back to her quickly. "That's big for me because I'm used to not only sending out messages at all times of the day and night but answering messages [at all times] of the day and night," she told me. "You know if I'm hearing from a boss, I'm going to [respond] right away, I don't care [if] I'm at dinner with my family—let me just rattle off really quick and respond." Yet Tonya hopes that if she can build a culture where people don't feel like they need to be working at all times of the day, this will empower them to take care of themselves and their loved ones. In this way, the ideas about resilience and wellness that she and her guests discuss on *Truth Be Told* are also being worked through in her own life and with her team. The hope is that as they work through these questions on air and off, there will be a domino effect and that culture will spread beyond the show's team.

Paradoxically, though, as we gain more power and status, we often stop paying attention to the views of others—even though our opportunities to connect with others actually grow. In other words, having the opportunity to make connections doesn't always translate into making the effort to connect. An exception to this default, Tonya makes an effort to connect with younger journalists and people outside of her organization in public radio. She works to understand their goals and is thoughtful about the advice she gives them and the message it sends to them about what's important. She talks with them about their purpose and values and encourages them to slow down so they can figure out what they actually love and hate within the work—not just what their bosses want from them—and to honor that as they make career decisions.

From a business perspective, the tendency to actively work toward understanding others' perspectives, and the ability to do so accurately, is a key leadership skill. In negotiations, being able to understand the other person's (or group's) perspective can also help reach better deals, and when people make the effort to take the other person's perspective, they can find win-win solutions that create value for both parties. But many leaders don't make the extra perspective-gathering effort that Tonya does, and many assistants take the perspective of their boss more often than the reverse. Why?

In research led by Keely Muscatell and Naomi Eisenberger, a team of us at UCLA explored whether people with different amounts of power and status use their brains differently when listening to other people's stories. In one study, we scanned volunteers' brains while showing them photos of other college students, as well as stories about those peers' experiences, thoughts, and feelings at different moments, like how they felt at the start of the new semester at college or while spending time with a friend. We also measured how students rated their own socioeconomic status (for instance, how much money, power, or prestige they had) relative to others on campus. Here, the students who felt like they had less status showed more activation in social relevance brain regions that help us understand other people's thoughts and feelings.

In a second study, we similarly found that when volunteers looked at the faces of peers expressing different emotions, those from higher socioeconomic backgrounds tended to show less response in social relevance and emotion processing regions to faces showing negative emotions. This suggests that without active effort, people who are higher in some types of social status may spontaneously take other people's perspectives less. Translated to a work context, those of us in leadership positions that come with resources and power may tend to have less of a sense of what is self- and socially relevant or valuable for our teams than others do, even though this is an essential leadership skill.

The good news is that we can develop skills to take other people's perspectives into account, and when people are prompted to consider other people's perspectives, it brings the social relevance system online. In research led by Columbia University's Adam Galinsky and colleagues at NYU, the University of Groningen, Lehigh University, and the University of Iowa, pairs of volunteers worked together to solve a murder mystery. Each partner received different pieces of information, and the correct suspect could only be identified by combining the information they received. But there was a twist: volunteers were randomly assigned to have different amounts of power—one member of each pair was randomly assigned to be the "boss" and the other the "employee." Moreover, half of the bosses and half of the employees were also specifically encouraged to try to understand the other person's perspective during the main task. Since the group to which each volunteer was assigned was random, their level of power (being the "boss") and motivation to take their partner's perspective (receiving the coaching) didn't have anything to do with their natural abilities or tendencies.

Adam and his team found that those assigned to have more power (the role of the boss) *and* to try to understand the other person's perspective shared more information with their partner, identified more of the unique clues only one partner had, and correctly identified the suspect more often. By contrast, when the boss wasn't assigned to understand their partner's perspective, the team had lower accuracy rates in identifying the suspect, even when the employee was trying to take the boss's perspective. This suggests the advantages of combining power and perspective taking. We can more intentionally work to understand where others are coming from, and doing so can lead to better team performance and collaboration. This is especially important to consider if you're a person in power.

However, it's also important to bear in mind that understanding someone else's perspective *accurately* isn't always as easy as it seems. For instance, my mom didn't realize my dad was shy until after twenty-five years together. Responding to my confusion over how she hadn't realized this for decades, she said, "I'm not shy and so I didn't know what shy was until now." Considering that none of us is gifted with the magical ability to read minds 100 percent accurately, we humans spend a lot of time trying to guess what others are thinking—or worse, assuming we already know. This is a natural inclination. Our brains allow us to simulate others' minds using our social relevance system. Sometimes we are right about what goes on in them, but often we are wrong, throwing off our ability to understand what matters to another person and therefore what they will do.

When we imagine what others are thinking (a process sometimes called perspective taking), we become more confident that we know what is going on in their heads. But since we are often wrong, the key is to take active steps to find out for sure. This might involve taking time to ask questions: How are you feeling? What do you prefer? What made you decide that? It might also mean taking time to read articles or listen to podcasts by people with different perspectives and identities. This opens the door for the kinds of communication that can help us mutually align our understanding of the world. Instead of going through life assuming we can guess what others are thinking and feeling, we need to actively seek out their perspectives to get it right. Listening to others' perspectives can also make them feel seen and improve our relationships with them.

## GROWING CURIOSITY

Working to understand where others are coming from—particularly when you have power—is key to being an effective leader, and as Roland and Tonya showed us, it can also help bring important new perspectives to solving problems. You might think you have to be an extrovert to do this kind of work. But my colleague Dani Bassett shows us that that is far from true—as does the research. The more core trait that we can grow in ourselves is not extraversion but *curiosity*.

Homeschooled as a child, Dani spent their mornings roaming ideas in books and nurturing kinship with trees and the natural world.

Today, Dani is a physicist by training, an innovative scholar, and a MacArthur Genius Award winner. Not only do they study curiosity and connection; they *live* what they study. They might not initially seem like a hub of social energy, since they speak quietly, judiciously, and are more likely to listen than interject in group conversations or team meetings; but for a wide range of topics that might come up in conversation, they know someone interested in, or working on, a related question. Like the networked roots of a tree, Dani is connected to many other scholars around the world. They make it part of their work to bring these people together in unexpected ways.

In their lab, Dani brings people together who might not otherwise converse. They run a training program where artists and scientists visualize networks in ways that advance both art and science, and they are constantly reading and engaging with people in a wide range of research areas. Like Roland and Tonya, Dani is generous in the connections they imagine between knowers, knowledge, and ways of knowing, and their research programs in domains ranging from material physics to neuroscience to sociology have been groundbreaking.

In their book, *Curious Minds*, Dani and their twin Perry Zurn invite readers to share the experience of curiosity and to expand our imagination about how different ideas and people might connect to one another. They encourage us to notice and nurture different kinds of curious practices in ourselves and others—maybe you like to go deep into an area and learn everything you can about it, maybe you flutter from topic to topic covering a broad ground, or maybe you are curious in a completely different way. They describe curiosity as "edgework" that connects dots in a network—linking bits of knowledge from different knowers, making meaning of it, and allowing us to "redraw the shape of our worlds, together."

In line with this view, as a journalist, Tonya Mosley foregrounds the expertise that comes from people's lived experiences as important and puts it in conversation with more widely recognized forms of expertise, as well as broader historical and political context. The result is new ways of understanding ourselves and the problems we care about.

In different ways, Roland prioritizes a wide range of expertise as well. He reads the work of physicists and materials scientists, but also psychologists and neuroscientists, architects and entomologists. He develops relationships with experts across sectors and asks a lot of questions, avoiding the assumption that he already knows how things (or people) work.

Likewise, we can purposefully become curious about people who might have expertise we've overlooked, and audit whose ideas are weighing most in our decisions. We can practice recognizing potential places where we also have the power to connect people and ideas and try on different kinds of curiosity if we want to. We can more purposefully go beyond small talk with the people we meet and interact with. Further, reading diverse genres, listening to podcasts hosted by people with a wide range of perspectives, watching films that evoke a wide range of emotions and experiences, and not limiting our conversation partners to people who speak our language will allow us to follow our curiosity deeper into the ground and higher into the sky, unbounded by whether someone lives nearby, or even in this time.

Connecting ideas can happen in all kinds of places with the right mindset, and our connections with others are some of the most profound sources of meaning and reward in our lives. Next, we'll learn more about these connections through recent neuroscience research about communication and influence. We'll see how the value and social relevance systems, both directly and indirectly, inform how we connect with others and how that has the power to change us and our environments.

# PART 3

# CONNECTION

# 7

# Getting in Sync

WHEN BRETT SUGGESTED WE *throw* our rings to each other on our wedding day, it was my turn to try out his New England eyebrow raise.

"It'd be awesome," he argued. "We could catch our own rings simultaneously!"

The suggestion wasn't completely without precedent: Brett really likes throwing me the car keys as I head out the door, or tossing me an orange from across the kitchen when I say I'm considering a snack. To catch the flying objects, it helps to know that he's planning to throw them, anticipate his move, and get my hand in the right place to respond. When I successfully catch the keys or the orange, it makes him feel like we are connected, anticipating and responding to one another's thoughts and needs. But *wedding rings?!* It seemed unlikely we could get good enough at this physical feat before the wedding, and wouldn't it be inauspicious to start off with a miss?

At the time, I found Brett's love of physical signs of synchrony funny, but didn't think much more about it—beyond the fact that we wouldn't be having a wedding day ring toss. I certainly hadn't considered the phenomenon from a neuroscientific angle, but a growing body of research in the years since has. In fact, recent findings suggest that Brett's interest in starting our marriage this way was actually pretty apt: when people's brains and bodies are more in sync, they tend to understand one another better.

It turns out that we're not alone in this—some nonhuman animals have this experience too. Neuroscientist Michael Yartsev's "bat cave" at UC Berkeley is home to about three hundred bats, hanging out together in clusters like teens at a high school dance. In the wild, bats are a highly social species that use sophisticated vocal communication, collaborate to build roosts, and huddle together in large groups to sleep. Like humans, they have squabbles over food and space and mates, but they also maintain long-term social relationships. In the lab, when pairs of bats are released in a "flight room" down the hall from the bat cave, the bats often choose to spend time together and tend to be active and rest at similar times. What's more, when Yartsev's team tracked the brain activation, vocalizations, flight patterns, and behavior of pairs of bats, they discovered that as the bats socialized, their patterns of brain activity became remarkably synchronized. The more the bats interacted, the more their brains synchronized; and the more their brains synchronized, the more the bats socialized with one another. Conversely, when pairs of bats were released in separate rooms, their brain activation was uncorrelated. In other words, the brain synchrony that the researchers observed wasn't simply about being active or sleepy on the same schedule; it was about social interaction.

Your intuition might tell you that if, like the bats, you were randomly paired with someone else for a few hours, there are some people you would be in natural sync with and others whom you would prefer to have been left in the flight room without you. It turns out that, like bats in the flight room together, human brains also synchronize with one another during social interactions. Recent neuroscience research suggests that this kind of brain synchrony may be one launchpad for successful communication.

Think about a time when someone explained a new idea to you that really "clicked" or a time when a friend understood you in a profound way. Maybe you've worked with a team that was great at anticipating one another's next moves and coordinating seamlessly or danced with a partner where it felt like you were moving as one. If you've ever found yourself with a friend, family member, or romantic partner whose sentences you can finish or who regularly turns to you with the same observation you were about to make, you might think, "Our brains work the same way"—and, sometimes, it's true! It turns out that the idea of being "in sync" is more than a metaphor and goes beyond coordination of our bodily movements. Shared *understanding* often involves two (or more) people's brains physically doing the same thing, mirroring one another's ups and downs, or at least coordinating in a way where one person's brain signals can be predicted from another's. And like the bats hanging out together in the flight room, brain-to-brain synchrony in humans is one foundation for learning from others. More broadly, brain-to-brain coordination (which includes synchrony) seems to be an important foundation for social communication and interaction.

In some ways, this might not come as a surprise—you may already feel "in sync" with some of the people closest to you, agreeing on things or having similar perspectives. Part of this might come from the process of synchronized activation of the social relevance system. As we saw earlier, arriving at similar preferences or ways of seeing the world can also activate the value system.

Research also suggests some ways that we can gain more agency in creating connections and achieving this type of synchrony, even with people we've never met before. This understanding can help us go beyond the idea that we naturally "click" with some people and not with others—that some people "get us" and others don't.

## HAPPY TOGETHER

Many of us know what it feels like to be "on the same page" or "in sync" with another person, either physically (dancing or playing music together, for example) or psychologically (feeling understood, seen, or coordinated on ideas). As Brett's love of kitchen-catch highlights, the simple act of sharing coordinated physical movements with another person can be rewarding. In research studies, being in rhythmic sync (versus out of sync) with others while drumming increases activation in the brain's value system, particularly for people who find it easy to get in sync with others. Likewise, people value being in sync with others while communicating, even when they don't get any useful information from it. A study led by University of Chicago psychologist Stephanie Cacioppo, for example, found that volunteers felt more positively toward and more connected with a partner when their nonverbal communication (specifically, rhythms of keyboard tapping while texting) was in sync—even though the partner was, in fact, a randomized, preprogrammed system and not a real person. This synchrony also activated the brain's social relevance and value systems, showing one way that nonverbal synchrony can tap into our social motivations.

When Brett throws me the keys and I catch them, our brains are temporarily in sync, each following the arc the keys take flying through the air, calculating where my hand needs to be to make contact. Research shows that it can be rewarding to be in sync with another person when we are working together (although our brains also synchronize when we go through painful and unpleasant experiences with others). For example, people show more activation in social relevance and reward regions when cooperating to jointly solve a maze than when solving it alone. The social relevance system helps us understand other people and brings us into alignment with them, and the value system rewards us when this alignment occurs.

But brain-to-brain synchrony isn't useful only for the intrinsic reward it provides—it is also a foundation for shared understanding that allows us to communicate successfully and work together. We often take it for granted that when we tell another person about some incident, they will be able to understand the experience we had, or that when we give a team member instructions, they will be able to understand what we have in mind for them to do. Brain-to-brain synchrony is part of what makes this possible, whereas a lack of brain-to-brain synchrony is associated with less shared understanding.

For instance, when people listened to a story told by a stranger, the more the listener's brain followed the same patterns of ups and downs as the speaker's, the more correctly they remembered what had happened in the story later. In other words, when the speaker's brain and listener's brains were in sync, the listeners learned from the speaker's story. In another study, the more students' brains mirrored the patterns of the teacher's brain, the more successfully students learned the facts from the teacher's lecture. The students whose brains showed the strongest coupling with the teacher also showed the biggest gains in learning scores. Likewise, in a laboratory experiment in which people were randomly assigned to solve problems alone or in teams of four, the more the teammates showed synchronized brain activity as they solved the problems, the better they performed. Research also shows that neural synchronization is key to successfully communicating information about experiences someone else might not have had yet.

You can probably imagine the wide-ranging benefits of the ability to bring our brains into synchrony and in turn learn from other people's experiences and knowledge. If your doctor understands your day-to-day schedule, they can better help make a plan that will work for you to remember to take your medications. If your boss can effectively describe her prior experience presenting to a client in a large, echoey boardroom, and you can imagine this room you've never been in, you can adjust your presentation so that the people at the far end of the table won't need to squint to read the type on your slides, and you can request a microphone so that all the team members can hear you better. Similarly, if you're working on a team where your communication brings you into sync with your teammates, you may find yourself being able to anticipate what your teammates will do, adding smoothness to the work you collaborate on.

## BRAINS OF A FEATHER

So far, we've seen that when two people's brains are in sync, they are more likely to understand each other. But it doesn't always feel like we are naturally in sync with the people around us, which may make it harder for us to connect with them. Maybe you've gotten into an argument with someone where you felt like they didn't understand the basic facts of the situation, or perhaps you have a coworker who always rubs you the wrong way and is oblivious to your point of view. To understand why we sometimes feel so out of sync (and, given the benefits of being in sync, how we might address it), it's helpful to look at why some of us are more in sync to begin with. Think of the friend who knows exactly what you need when you've had a rough day, the sibling who gets all your family dynamics, or the coworker who shares your sense of humor. What makes some people "just click"?

Consistent with the intuition that "birds of a feather flock together," research led by Carolyn Parkinson, Thalia Wheatley, and Adam Kleinbaum, at Dartmouth College, showed that the brain patterns of close friends are remarkably similar to one another when making sense of the world. The team had a cohort of MBA students watch short video clips on a wide range of topics—from an astronaut in space, to baby sloths in their sanctuary, to a review of Google Glass, to slapstick humor from *America's Funniest Home Videos*. Close friends within the group showed more synchronized brain responses to the clips, whereas those who were friends of friends, or even friends of friends of friends, showed less similar responses. In parallel, students who responded more similarly to the videos at the start of the semester were more likely to become friends as the year progressed. This suggests that people whose brains work more similarly may also find it easier to click—or, as Carolyn says, "brains of a feather flock together."

Led by Ryan Hyon, Carolyn's team also found that even in the absence of an external stimulus like a video clip, when people let their minds wander wherever it may take them, friends show more similar brain patterns to each other than to people they aren't close to. And it wasn't only American college students—the team collaborated with Yoosik Youm and Junsol Kim and found similar patterns in residents of a Korean fishing village. Villagers who were closer together in their social networks (a friend versus a friend of a friend of a friend) showed more similar patterns of brain activity while their minds wandered. They also found that those who lived physically closer to one another and presumably shared more similar day-to-day experiences also showed more synchrony in the brain. Together, this research suggests that the brains of people who are friends and of people who share similar life experiences (whether they like each other or not) tend to respond to the world in more similar ways.

But our day-to-day lives aren't only created by friend groups and physical proximity—media shapes much of our lived reality and thus our "natural" ability to connect with each other. This works in reverse, too: our current attitudes shape which media and conversation partners we seek out. For instance, you may have noticed that a friend who primarily gets her news from Fox will be exposed to stories about different topics and will see different angles highlighted for the same topic, compared with someone watching CNN. Importantly, when people watch and listen to the same media, spend time together, and communicate, their brains and bodies synchronize. By contrast, people who consume different media and have different life experiences diverge in how their brains respond to identical content. The media we consume—and how it portrays stories—can influence how we see the world and interact with others on topics ranging from violence to gender, sexuality, race, and political preferences (for example, about the justice system or immigration). As such, media institutions play an important role in cultivating our collective worldviews, and our background knowledge and assumptions shape our brain responses to media.

Recent research highlights how consuming the same media brings people's brain responses into sync with one another. For example, in one study led by Uri Hasson at Princeton University, volunteers watched half an hour of Sergio Leone's classic film *The Good, the Bad and the Ugly*. Their brains were scanned as they watched the young Clint Eastwood, as a bounty hunter, staring down enemies in gunfights, searching for gold. Leone's cinematography builds tension, alternating between sweeping panoramas, steely close-ups of a gunslinger's eyes, and a hand reaching for a gun. Impressively, brain activity in one audience member as these tense scenes unfolded could predict brain activity in others.

Some of this synchrony was driven by the sensory experiences the movie created, which were objectively similar across volunteers, since they were watching the same movie. However, the audience also showed remarkable synchrony in brain regions that track higher-order thinking, like self- and social relevance. This doesn't necessarily imply that the volunteers were having the exact same thoughts; rather, it's likely that shared background knowledge about pop culture, societal values, and norms generates overlap in which parts of a film or story people find socially relevant, and vice versa. Some of this synchrony might also be attributable to people's social relevance systems anticipating what would happen next, similar to what we saw earlier with the kids who watched *Partly Cloudy*. In a close-up shot of Clint Eastwood's face, as he prepares for a duel, we can imagine his thoughts, collectively transported into the dusty scene. These same bits of background knowledge are also accessible (and likely incredibly useful) to the various creatives involved in making media, from writers to directors and beyond.

Media also shape our brain responses to political issues, in conjunction with the partisan identities we bring to the table. When people from different political parties consume different news (for example, watching Fox versus CNN), our brain responses become more similar to the people who are engaging with the same media and ideas as we are (Democrats show more similar brain responses to other Democrats, and Republicans show more similar brain responses to other Republicans). In one research study, emotional words in news content elicited more similar responses in people with similar partisan identities, compared with those with different identities. When people have exposure to the same media diets, and when they share experiences, they show more similar brain responses.

This influence is likely bidirectional. Just as we gravitate toward different content because of who we are and what we believe, changing the content we consume can reshape what we think about and in turn some of what we think is important and believe. When, in fall 2020, researchers incentivized a group of Fox News viewers to watch CNN (at times when they typically watched Fox), it changed not only the topics they were knowledgeable about (increasing their knowledge of topics typically covered on CNN and decreasing their knowledge of those typically covered on Fox) but also what issues they thought were important, what they thought of different policies, and even how they evaluated different politicians, relative to a control group of Fox viewers who kept watching their typical news. In other words, when the researchers changed people's media diet, it changed their beliefs, opinions, and what they thought were important issues.

Knowing this, we can consider what kinds of media we are consuming ourselves and how what we consume might be shaping our views and values. And if media content can change what we think about, it follows that it could, by extension, open up new possibilities for whom we feel in sync with—and whom we *don't*. After all, media are a big driver of assumptions we make about others, and these assumptions can pull us in or out of sync.

### GREEN MY EYES

It's late at night when Lee gets a phone call from his friend Arthur. Arthur has just gotten home from a party where he lost track of his wife, Joanie. Arthur suspects she is cheating on him: "You can't trust her," he says. "I swear to God. I swear to God you can't."

Lee tries to soothe Arthur from the other end of the phone line, and there is a woman in his bed, listening in. We don't know much about her.

In fact, in the story, "Pretty Mouth and Green My Eyes," by J. D. Salinger, much is left to the imagination. Is the woman in the bed the missing Joanie? We can't be certain, but some of the assumptions you might make about the characters would depend on what you brought *to* the story. What else do you know about these people? About this situation?

Research led by Princeton neuroscientists Yaara Yeshurun and Uri Hasson highlights that as our starting assumptions diverge, our starting context biases the way our brains respond to what comes next. By the same token, creating shared context biases our brains to see stories similarly to other group members, as though we've been bats flying around together. In Yaara and Uri's research, volunteers had their brains scanned while they were given one of two different backstories before listening to an adapted excerpt from "Pretty Mouth and Green My Eyes." The researchers gave half the volunteers a backstory that makes it clear that the woman in Lee's bed is Arthur's wife, Joanie. The researchers gave the other half of the volunteers a different backstory, in which Lee and Joanie are not having an affair, and Arthur is simply paranoid.

When listening to the excerpt from Salinger's story, volunteers who were given the same backstories interpreted Arthur and Lee's conversation similarly, and their brains were in sync. In contrast, when people were given different backstories, even though they were responding to *the same words in the next part of the story*, their brain responses diverged. In other words, creating shared context biases our brains to see stories in similar ways to other group members.

The research team manufactured a situation in which one group shared one set of beliefs (they *knew* Lee and Joanie were having an affair) and the other group another (they *knew* Lee and Joanie were innocent). But the beliefs that we have coming into conversations can feel equally strong in our daily lives. One reason we might get off on the wrong foot with someone—or even why our opinion of them may shift mid-relationship—is that incorrect assumptions pull us out of sync. Even in cases where we share the same basic values and politics and watch the same news shows, there are many other forces that shape our assumptions about the world and our social interactions. This is true in large-scale political discussions, but it's also true in small exchanges at home or around the office.

Imagine that Joyce has been working on a presentation for several days and is nervous but excited to get feedback from her manager, Maya, whom she really looks up to. Maya is excited too. She thinks of Joyce as a superstar and is hoping to help her move her ideas forward in the company. Unfortunately, the night before the presentation, Maya doesn't sleep well, rushes through her morning in a fog, and still hasn't had lunch by the time the presentation rolls around in the early afternoon. As Joyce moves through her presentation slides, she can't help but focus on Maya—not only is she Joyce's manager, but her flat, bored expression is impossible to miss. Joyce keeps her cool and picks up the pace, but after the presentation, Maya is short with her.

"Interesting approach," she says. "I wonder if Cheryl might also have some useful feedback." Then she hurries from the room.

Joyce is crushed.

But Maya really did think the approach was interesting—she was impressed with the presentation, excited to work with Joyce, and imagines Cheryl will be as well. She's also *hungry* and needs to grab a bite to eat to make it through the rest of her afternoon meetings.

If Joyce could see Maya's hunger or had interpreted her blank expression as fatigue instead of boredom, then she could adjust for this in her interpretation of the feedback. But she doesn't have this information. With the wrong initial assumption, she can't make a good prediction about where Maya's thoughts and actions are headed. So as

Maya scarfs down lunch and shoots off enthusiastic emails to higher-ups, including Cheryl, Joyce is back in her office, defeatedly moving through her afternoon routines. They are out of sync.

Here, neuroscience research illuminates not only the idea that people with different knowledge, experiences, and assumptions diverge, but also *how* and *why* they might diverge, even when they are hearing the same words or objectively seeing the same facts. Since Joyce doesn't know that Maya is hungry and tired, her brain interprets the ambiguous "interesting approach" differently than Maya meant it.

Conversely, when people share assumptions about the backstory, facts, or context of a given situation, their brains are more likely to interpret new information in parallel ways. If Joyce knew that Maya was thinking about the related project that Cheryl had recently pitched, she would understand why Maya suggested they talk, and Joyce and Maya's social relevance systems could be in closer sync. Likewise, in the political sphere, people from different groups often wildly overestimate the extent to which the other group dislikes and even dehumanizes them, and correcting these assumptions reduces harmful behaviors associated with them.

This is part of what motivated Tonya Mosley to change the way she engages as a reporter. After years covering high-stakes, emotional, political events, Tonya told me, "I could hear, over and over again, that there was not a meeting of the minds on some basic principles of humanity, about listening to each other, caring for each other." Being in sync requires a common ground and context to depart from. But getting there is a whole other thing.

Since it seems impossible not to have assumptions about what's going on in the world around us, the broader situation could seem doomed: We can't always be right, but we will always make assumptions. But the fact that people have hidden, and wrong, underlying assumptions about what the other side thinks suggests possible inroads for improving communication. Simply *knowing* this, and keeping it front of mind, puts a flashlight in your hand.

But how do you shine it around? How do you ask: *Where are the possibilities?*

One way to do this is to have conversations that explicitly surface background assumptions and interpretations. This approach is consistent with evidence from neuroimaging research. Conversations can bring people's brain activity into alignment during the conversation itself, and the effects of being aligned persist later as well. In one study led by Beau Sievers, Thalia Wheatley, and Adam Kleinbaum, at Dartmouth College, volunteers watched movie clips that could be interpreted in different ways before discussing what had happened with others in a group. Having the opportunity to discuss their different interpretations brought group members' brains into greater alignment when they watched the clip again. Importantly, it also resulted in greater alignment when they saw new clips that hadn't been discussed, suggesting that developing shared understanding can help get our brains in sync more generally, including when making sense of new evidence and situations.

In this way, if Maya and Joyce had eaten a quick lunch together after the presentation and had more time to touch base, Joyce may have learned information (like how hungry Maya was) that would have helped her understand her manager's comments. Or if Maya had shared a quick caveat about her morning with Joyce before the presentation and reassured her, Joyce might have felt better. A lot of informal communication is necessary to understand one another's thoughts, feelings, and differing perspectives. This is true both when we come from different starting assumptions and when our assumptions unexpectedly diverge.

The Dartmouth team's research also highlights that, across the groups they studied, groups that had someone whom raters perceived as "higher in status" often looked like they had adopted a consensus view, when in fact something else was going on. These high-status folks steered the group; they made strong declarative statements, exuded confidence, cut people off, questioned others' positions, and generally pushed others to adopt their viewpoints. If we only looked at the survey data, it would seem they succeeded: group members followed their lead behaviorally. But appearances can be deceiving. When the research team scanned everyone's brain again, they found that the group members were, in fact, out of sync. The lower-status group members who had been pushed to agree were not actually aligned—they were simply going along to get along. Outside the lab, Adam Kleinbaum says, he can imagine that this is the scenario where people might agree with their boss in the meeting but then go back to their desks and half-heartedly work on the agreed-upon plan or maybe even undermine it.

But when a group had someone who was liked by and connected to many peers (more central but not necessarily "high-status" in the social network), the group members agreed in both their verbal responses *and* their subsequent brain activity. These more central folks were more deeply influential, but also more flexible in their own positions—they were happy to move toward the opinions of others in the group. They asked follow-up questions, asked people to explain what they meant, and their groups ended up more in sync.

This is the curiosity that Tonya models with her guests on air—and a way we can each practice listening more openly. We can ask questions we're genuinely curious to learn the answers to and be open to the idea that we are capable of change. It matters who is in the group not only for the information they bring to the table but also for the group's ability to arrive at a real consensus. Would you rather be the person who bulldozes others into a false consensus or who creates deeper neural alignment? In this vein, Tonya sees the conversations she has with people on air as opening up new paths toward a more functional society, for how we might (re)imagine solving problems together.

If we do it right, conversations offer one way that we can surface starting assumptions, give others a chance to share their perspectives, and get in sync. But in what other ways might understanding synchrony help us set ourselves up to build a stronger sense of connection and, perhaps, shared perspective with others?

## EXPANDING AND EXPLORING

Having seen how shared neural representations go hand in hand with shared thoughts or understanding and how synchrony can serve as a starting point to successful interaction in both our social and professional lives, we might be tempted to conclude that increasing the amount of synchrony we experience with others should be the main goal. Yet—perhaps counterintuitively—if we want to learn from others and explore new ways to see the world, we may not want to be perfectly in sync, or on common ground, all the time. Instead, sometimes we might want to aim for a different form of coordination, in which we complement one another and push each other to explore new territory.

As we've seen, there is value to the ways we are different, the ways we *don't* sync up. Surprise captures people's attention, and people like fast-paced, deep conversations where they learn new things and talk about novel ideas. To grow in our sense of possibility, we want to connect ideas that haven't been connected before and engage with people whose life experiences might suggest different ways of interpreting the same evidence we are sure we understand.

If our brains were fully in sync with one another all the time, there would be little room to discover new ideas or explore the wider landscape that makes the best conversations fun. So instead of aiming for continuous synchronicity—like only spending time with people with whom you immediately "click" or avoiding any topics you think might seem far afield, personal, or contentious—maybe the goal should be to get in sync enough that we can depart from the same common ground, then diverge in where we guide the conversation, facilitating exploration of a wider ground or deeper connection.

A recent study that I collaborated on with a team at Princeton highlights the conversational benefits of divergence. Led by Sebastian Speer, Lily Tsoi, and Diana Tamir, along with Shannon Burns and Laetitia Mwilambwe-Tshilobo, we scanned the brains of pairs of either friends or strangers as they moved through a simple game called "Fast Friends." The game—designed to forge friendships between strangers for lab studies—involves taking turns asking and answering questions that build in their level of intimacy, starting with ice breakers like "Given the choice of anyone in the world, whom would you want as a dinner guest?" and "What would constitute the perfect day for you?" The questions eventually build to more personal questions like "Is there something that you've dreamed of doing for a long time? Why haven't you done it?" and "If you were to die this evening with no opportunity to communicate with anyone, what would you most regret not having told someone? Why haven't you told them yet?" and finally, "Share a personal problem and ask your partner's advice on how they might handle it. Also, ask your partner to reflect back to you how you seem to be feeling about the problem you have chosen." When people play this game, they tend to enjoy it and feel closer at the end—even if they had no prior connection or have different attitudes about important issues, and regardless of whether they expect to like one another.*

During our volunteers' conversations, we observed an interesting pattern in their social relevance systems. As we expected, pairs of strangers increased their synchrony over the course of the conversation—they built common ground. Close friends, on the other hand, started more in sync, but then their brains diverged over the conversation.

---

* One caveat here is that when introverts play together with no instruction about the goal of the game, they end up feeling less close to one another than when extroverts play with one another. That said, when people are told that the goal of the game is to become close to one another, the differences between introverts and extroverts go away—in other words, giving people the goal to connect can overcome differences in their natural tendency to try to connect with strangers.

In these intimacy-building activities, we found that the friends (who started closer to one another than strangers) tended to cover a wider range of topics, diverging into new territory and switching back and forth between topics. The strangers tended to bounce back and forth less, sticking more closely to a smaller range of topics.

Importantly, although friends' brains were more likely to diverge over the conversation, the strangers who covered many topics and wide conversational ground also showed more divergence in their brain responses over time and reported enjoying the conversations more. This means that when strangers' brain responses looked more like friends', they also had better conversations. Establishing common ground early, and then diverging, was associated with having a better conversation.

I think about this research sometimes when I'm in a situation with people I don't know very well. Dartmouth psychologist Emma Templeton has found that strangers will use a more predictable set of what she calls "launchpad topics" like the weather or asking "where are you from" to get into a conversation, whereas people with more established friendships can dive right into deeper and more varied topics—and often have more fun. Instead of spending the whole conversation discussing the weather at a birthday party for one of my kids' friends, after establishing that basic common ground, I might ask a fellow parent whom they would invite to the party if they could have anyone else in the world as a guest, or what they've dreamed of doing that is challenging to do while juggling kids. Their answers often surprise me and tend to shake up assumptions I may have had about them.

More broadly, the findings from our study are consistent with research described above on how bringing together different ideas can spark innovation and how drawing on a wide range of ideas makes us more interesting conversation partners. One of the key members of our team, Shannon Burns, has pioneered the idea that during successful communications, brains should complement one another, and she brought that perspective to our work on conversations. In a follow-up study in which strangers negotiated about how to distribute a budget to address problems related to education debt and the environment, Sebastian, Shannon, and the team discovered that dyads that explored more ground had better conversations and eventually came to closer consensus about how to solve the problems.

Although there's a lot more work to be done here, this research highlights some of the benefits of already having established common ground. Once we are secure in that synchronized foundation, we can explore new topics and learn new things about ourselves, others, and the world. At the extreme, we don't want everyone to have the exact same neural response all the time. That would lead to boring conversations and hinder creative thinking. Instead, in many cases, we want to coordinate in more sophisticated ways—dividing mental work, taking turns leading and following in conversations, and exploring new ground.

What about when we feel hopelessly out of sync? Being listened to and heard can depolarize people's attitudes by decreasing defensiveness and increasing productive self-reflection. The reverse is also true: we can benefit from listening with the goal of understanding others' points of view—whether that is to allow a neighbor with differing views to feel heard or seen, to gain valuable knowledge, or to be a better leader. In Sebastian and Shannon's negotiation study, when we nudged people to compromise, rather than persuade their conversation partner, they explored a wider range of ideas and ended up more aligned in their decisions at the end of the conversation. For those from whom we know we differ, exploring what we might have in common—and remembering that we don't have to land on the same page for the conversation to be valuable—can support constructive connection.

Finally, although it can sometimes feel effortless to be in sync with the people closest to us (like when I successfully catch the car keys Brett throws across the room), it takes deliberate work to set up the conditions to get there—asking questions about how someone's day is going, what led to their recommendation, and then being present when they answer. (Just leave your phone upstairs, Emily.) So much of this groundwork for how we can come into sync—or more broadly, into coordination—has to do with how we share information and stories with each other. This hits on many levels, from the media we choose to consume that shape our collective consciousness to the happenings in our everyday lives. All of this informs how we see the world and connect (or fail to connect) with each other.

But what motivates us to share in the first place? Understanding this might help us have more agency in what we decide to share, how we connect with others, and in understanding how others might tap into these same motivations for their own ends.

# 8

# Small Acts of Sharing

ONE DAY, SWIPING PAST yet another dating profile that just wasn't right, Eric landed on Laura's profile. From the preview, he could see that she was beautiful and had an impressive set of intellectual interests. He tapped the photo to open the rest of the profile. There was a picture of her in a lab coat but also a picture of her in an '80s-style workout class, her face aglow with a level of enthusiasm greater than all the other participants' combined. While people who "don't take themselves too seriously" have become a dating app cliché, Laura clearly didn't. She was (potentially) all the things he had been searching for. He paid a few extra dollars on the app to send her a virtual rose, and shortly after, my little brother was forwarding me a voicemail she had left him in her dreamy British accent, agreeing with my kids' excitement about cherry tomatoes ("a good tomato [*which she pronounced toe-mah-toe*] is a thing of beauty," she murmured into the phone). I swooned.

Why was Eric forwarding me voicemails from a woman he had just started talking to? you might ask. I suspect he couldn't contain his excitement about her, felt proud to be talking with her, and wanted to share that feeling. He also knew that I, too, revered a good tomato. "You should probably marry her right now," I joked.*

Although the high of meeting someone perfect for you comes along only once in a while, these small acts of sharing are incredibly common—and they are important building blocks of our relationships with one another. If you think back over the last twenty-four hours, chances are you, too, have shared with others. Maybe you told a friend about your day, shared your expertise with a coworker, posted an update about your life on social media, showed off your adorable cat, or shared a highlight of your vacation where you looked particularly good (not that *you'd* ever do that to make others jealous, but *some* people might). Or maybe it was more serious. Maybe you felt overwhelmed seeing images of wildfires in the West, flooding and devastation in coastal cities, or stories about violent conflict in the world. You called a friend to talk about it, or maybe you shared a news article that outraged you along with a hot take. Maybe you didn't actually feel outrage, but wanted to signal to others that you are the kind of person who cares about a particular issue.

Perhaps it was a quieter, more intimate kind of sharing. On my parents' wedding anniversary, over dinner, my mom, grandma, and I shared memories of my dad, who passed away a few years ago. My mom told me a story I hadn't heard before about the songs he'd sing on her voicemail, recalling my father's love of music and how there was always music in the house. How he'd take her in his arms and they'd dance in the kitchen or, at parties, the times they danced the night away. I remembered how proud my dad was of my mom's poetry and how much fun they had dancing at my friend Alex's bar mitzvah. My grandmother smiled, remembering the gardens he tilled and planted and how, in addition to the usual garden fare, he'd surprise

---

* Okay, if we're being real, I was only partly joking. I actually thought maybe she was an MI6 agent, trained by an expensive team of government agents to perfectly play the role of my brother's dream girl, since she seemed so darn competent along so many dimensions.

my mother with plants she especially loved, like nasturtium, aspara-
gus, all different kinds of salad greens. My mom recalled how he loved
to share the things he loved and how one year when my mother asked
what he wanted for his birthday, he said that he wanted her to let him
prepare all her meals for two weeks, to read articles he'd provide, and
to watch five documentaries with him. As I listened and shared my
own memories, I felt closer not only to my dad, but to them.

Sharing with others is one way we bond with them and feel less
alone, and it helps us make sense of our own experiences. When we
share, we also make implicit and explicit decisions about how to present ourselves to others ("check out my awesome girlfriend's voice")
and also guess about what others will think of what we share ("my sister
will love this"). On a larger scale, these decisions about what to share
can contribute to broader cultural norms and change. Sharing shapes
what people think is true, what their preferences are, how political
movements unfold, and what people care about in popular culture.

To be sure, some of the same kinds of motives—like feeling connected to others, being part of a group, or having power and status—
can lead to much more harmful effects online. Our desire to connect
and share can be exploited to keep us scrolling, clicking, and engaged
on online platforms, as well as to change our preferences and behaviors, and maybe our brains. These forces can also lead to extremism
and bullying, successful scams, the dumpster fire of hate and outrage that stoke real-world violence, and the ability of political actors,
marketers, and trolls to sway our opinions in ways that undermine
democracy. An in-depth discussion of the network mechanisms of
information spread, the social, financial, and political interests that
fuel sharing online, and how the Internet and various platforms shape
these forces is beyond our scope here,* but it is an area of active

---

* In brief, content that gets shared widely online is now heavily shaped by algorithms that social media companies and other online platforms have developed to
curate what we see and ultimately what people share. The motives of these companies are often not the same as those of the individuals who are being targeted by

research, and a growing body of neuroscience research has begun to focus on the darker contexts where more harmful sharing unfolds, including the effects of false information and radicalization. Research on the neuroscience of sharing is still in its infancy; nearly all the studies we will explore in this chapter focus on people sharing ideas or information about themselves or about topics like health and the environment written by trusted news sources. We started doing this research before the major rise of social media as a primary information source and before the digital disinformation campaigns that have become a major public health concern. Though much of the neuroscience research so far has focused on the benefits of sharing for individuals (for example, bonding with people we know and building relationships or spreading messages designed to motivate prosocial or healthy behaviors), both the positive and negative implications of sharing make it important to understand what makes people want to share in the first place. This understanding might allow us to become more aware of how we can derive the benefits that come from connecting with others and also how this can be exploited by people who don't share our interests.

As with so many decisions, what we choose to share comes back to a value calculation, with the self-relevance and social relevance systems playing key roles. And the things we share influence ourselves and others—though not always in the ways we expect.

## WHAT MAKES US SHARE?

I first became interested in what makes people want to share certain ideas with others when I was a graduate student in Los Angeles. While my work was focused on the neuroscience of health campaigns, it

---

ad-driven business models. Interested readers might check out *The Hype Machine*, by Sinan Aral, for more on how platforms shape what we see and share online.

seemed like everyone around me was tied to the entertainment industry. Billboards lining the streets touted the newest movies and TV
shows. Many of my friends worked in the industry as assistants to
Hollywood producers, set designers, or assistant editors.

At weekly game nights, I often heard from them about crazy bosses
who might throw a plant at you if you were late bringing them their
favorite protein bar, as well as about the people reading new "spec
scripts" in the "slush pile." Their job was to triage unsolicited submis-
sions and write "coverage" for their bosses to review. The decisions
they made about which film ideas to promote to their bosses would
influence what would become popular culture and ultimately what
large numbers of other people would think was cool, relevant, and
worthy of attention.

Meanwhile, I was trying to figure out how to design more effective
health campaigns. My PhD adviser, Matt Lieberman, and I believed
such campaigns could be more powerful if the people they directly
convinced to change their behaviors also convinced their friends to
join in. Indeed, research has shown this to be true both for the health
messages we were interested in and for a wide range of other messages
as well. But of course, people don't tell their friends about every new
idea they come across. We wanted to understand what motivates people to share some, but not other, pieces of information, and what goes
on in the brain when they make these decisions. Then I thought about
my friends wading through the slush pile. Voila: a study was born.

We designed an experiment in which two groups of undergraduates
at UCLA played interns and producers at a television studio evaluat-
ing pitches for new shows. Our team started by gathering and coming
up with about two dozen ideas for possible TV shows. One, *Mafia*,
centered on two best friends—one crafty and wise, the other a skilled
gunman—who try to rise to the top of a mafia family. Eventually, the
crafty one kills his best friend and partner to become the new head
of the crime family. Another, *Athletic Adventures*, was a college comedy about the hijinks of a small sports team that nobody really pays

attention to and that gets away with a lot of mischief as a result. Dan, the star player, appears to be a nice, well-adjusted guy, but becomes a completely different person off the court. On the one hand, we probably weren't going to win Emmys with these ideas, but on the other hand, looking back at what has gotten made in the decade since, I feel like we weren't far off . . . though maybe we should have just based a show on our lab, since *The Big Bang Theory* was such a hit.*

After we had our pitches, we gathered our volunteers and monitored their brain activity as they were asked to imagine that they were interns in charge of deciding which of the ideas to pitch to their boss, the producer. They watched a slideshow of our twenty-four pitches, which included a title and a few sentences about each show along with an image you might see promoting the show. After seeing each slide, the participants rated how likely they would be to pass the ideas on.

Once they were out of the brain scanner, they recorded a video in which they pitched the shows they thought had potential and critiqued the ones they thought would flop. Some folks really got into it, beginning to imagine themselves watching the shows or how the shows might be positioned in the market. In the next phase of the study, we showed these videos to a second group of participants, the "producers," who rated *their* intentions to pass on the information to someone else. This allowed us not only to get some feedback about which of us scientists might be ready for an alternative career in Hollywood, but more important, to observe what was happening in the brain when the interns initially decided what to share, and then to determine whether this brain activity tracked which ideas spread successfully from the interns to the producers and beyond.

At this point, it probably won't surprise you to learn that we found that activity in the medial prefrontal cortex and posterior cingulate

---

* Thanks to real-life TV show writer Emma Fletcher for this excellent suggestion, which I interpret as confidence in the exciting nature of our research.

cortex (regions involved in self-relevance, social-relevance, and value processes) of an "intern's" brain in response to an idea correlated with how likely they were to decide to recommend it to their producer. But critically, higher activation of key regions in the self-relevance, social-relevance, and value systems in the "interns" also predicted how likely the "producers" were to choose to recommend the ideas later as well.

What do these results tell us about why people share? At one level, although brain scans can't tell us exactly what people are thinking, considerations about self-relevance might include evaluations of whether the idea is relevant to me, or more generally what I think and feel about it. Considerations of social relevance might include whether the idea is relevant to the people I'm thinking of sharing with, or more generally the people around me, and what *they* might think of the content. At the intersection of self- and other-focus are thoughts about what it will say about me if I share it with others (remember my worries about sharing my musical tastes with my friends?).

At another level, I also think these findings are consistent with a deeper function of sharing, one that transcends the small decisions we make about whether to forward a meme or pilot idea to our boss. They speak more directly to our fundamental, human needs to feel good about ourselves, connect with others, and maintain social status. In other words, our small, daily choices to share can tap into deeper needs that we may not even be conscious of.

## WHAT DOES THIS SAY ABOUT ME?

Three days and one video-date-cookie-baking-session after Eric sent Laura the virtual rose, Eric's mind was made up. "I wanted to lock it down," he told me. He hadn't yet met Laura in person, since it was the height of the COVID-19 pandemic, so it felt extra important to plan the perfect virtual date. That's when he remembered the decks of cards I left at his house a year before.

Back then, he and a friend had helped me prepare a workshop for a group of business leaders. As part of the workshop, I shared research about what happens in the brain when people have a chance to share information about themselves with someone else, and I also had the participants work with this deck of cards, each card with a different question printed on one side. It was—you guessed it—"Fast Friends." Now Eric flipped through the cards with Laura in mind.

"He started with the easy questions," Laura told me. But then they got to the deeper questions in the third deck. "There was a question, something like, 'If you die tonight, what is something you'd be sad you hadn't said?'" Laura laughed, recounting the date. "I thought, 'I think I'm going to fall in love with this person,' but I didn't say that at the time." Eric's plan worked, and soon the two met up in person.

Importantly, it isn't only newly dating couples who appreciate the chance to share things about themselves. I remember the first time my kid, Theo, snuggled up on my lap and asked me, "How was your day?" I smiled and told him about walking to work that morning past spring magnolia trees beginning to bloom, about a meeting I'd had with a student who had had a rough few weeks, and about a talk I'd given on Zoom to share my research with colleagues in other states.

He listened, patiently and thoughtfully, and when I was finished, he announced: "We have a kindness bingo chart in our classroom. Asking 'How are you?' or 'How was your day?' gets you a square."

I laughed. Did asking these questions really constitute an act of kindness? But after some thinking, I realized it *did* feel good to tell him about my day—he seemed so genuinely curious about the answer. I was reminded of bedtime stories my parents used to tell me about their own childhoods. I had loved hearing about mundane details like my dad eating hot dogs by a stream or my mom considering whether to throw banana slices out of her family's apartment window when she was a kid.* My parents liked sharing those stories too.

Whether in conversations with friends, on dates, or at work, you've probably shared stories and details about yourself thousands of times. I feel comfortable assuming this, because humans talk about themselves. A lot. Why do we do this?

When Theo asked how my day was, I felt good telling him, and it turns out that I'm not alone. Princeton psychologist Diana Tamir has found that people find it inherently rewarding to share information about themselves with others. When her team looked at the brain activity of volunteers choosing whether to share information about themselves or about other topics, the volunteers showed greater activity in the brain's value system when disclosing information about themselves. In fact, when given the choice to share information about themselves or to share information by answering trivia questions—and to receive different amounts of money, depending on their decision—the volunteers were willing to forgo money by choosing to share information about themselves, even when they could have earned more by answering the trivia questions. Sharing information about ourselves is its own reward.

This may seem like the epitome of selfishness, but there are good reasons why our brains value self-disclosure. For one, sharing about our preferences also allows us to coordinate better with others. If I like the pointy bite of the pizza and you like the crust, we should make sure to discuss that so we don't end up trying to be nice to one another and each getting the parts we like least. In real life, this was another reason I came to really root for Laura. She likes the pie crust, and I love filling. That's a match come the holidays.

---

* I recently asked my mom why she did this. She says she doesn't know why. She wondered what would happen, and as a kid, I have to say, it made total sense to me that if you lived in an apartment building high above the ground and had access to a banana and a chunk of time when an adult wasn't watching, you might want to see what happens if you throw a slice out the window and let it splat on the pavement.

Another reason self-disclosure is powerful is that in day-to-day life, sharing about ourselves is a key way to strengthen our bonds with other people—it's hard to feel close to someone you know nothing about. Although it feels natural that we share more about ourselves with people we like, experiments show that it sometimes goes the other way too. Disclosing personal information makes us feel closer to the person we are sharing with, and people tend to like others who disclose personal information to them (within some appropriate limits).

This is part of why "Fast Friends" works so well, but it's worth noting that this personal information doesn't have to be strictly "personal." Sharing ideas, news, memes, and other information that might not be directly about us is another way we tell people who we are and what we care about. This helps us understand some of the harmful behavior we see online (such as polarization, people falling prey to extremism), but I also think of my dad once emailing me and other friends and family to explain why he had gone on a ten-day meditation retreat and including some YouTube videos about meditation. Or Eric forwarding me Laura's voicemail about tomatoes. Or me sharing my music playlist with my friends—I was nervous because it felt like an expression of who I am.

Understanding sharing as a form of self-expression—which, as we've seen, humans are inherently motivated to do—helps us understand why people amplify certain kinds of messages online. In a set of studies led by Dani Cosme, Christin Scholz, and Hang-Yee Chan, my team investigated how we might encourage people to share high-quality news articles or social media posts that promote healthy lifestyles, discuss climate change, or call for people to vote in political elections. If many people shared high-quality information in these domains, we reasoned, we could potentially shift norms and make it easier for people to make healthy choices for themselves and for their communities. In these studies, we recruited people to participate in a study online. Those who signed up were randomized to write brief posts that either highlighted why certain messages were relevant to them or that simply described what the message was about. Then they rated how relevant the message was to them and to others and whether they'd be willing to post it online.

It wasn't very surprising that when people told us that the news articles and social media posts initially felt personally relevant to them, they were also more likely to want to share the information with others. What was more interesting was what happened when we asked people to write a brief description about how a given news article or social media post might be personally relevant to them. "My family's home was affected by a recent hurricane," wrote one participant. "Climate change could be affecting more people and more severely now and in the future." This simple act—connecting the information with their self-concepts and personal experience— made them want to share the articles with others more than when they only thought about the basic content of the articles. In a series of follow-up studies, we found that the more people rated news articles as relevant to themselves, the more activation they showed in the self-relevance system, and the more likely they were to want to share the articles online.

We're more likely to share information that we perceive as self-relevant, and simple prompts that invite people to write about why content is important or relevant to *them*—asking them to share why *they* care about an idea or cause, in addition to the idea or cause itself—can increase their motivation to share. As we've seen, people are eager (and even willing to effectively pay!) to share information about themselves with others. So if there's an idea you're hoping others will want to talk about and share—whether it's health advice from a trusted source or a new product you're excited about—these kinds of prompts might encourage others to help amplify it, and get something valuable out of it themselves. You might also pause when others use similar tactics to draw your attention and encourage you to share things that aren't from reputable sources or that might stoke political divides or encourage bullying.

## LOOKING SMART, COOL, AND CAPABLE

Sharing is an inherently social activity, so it's intuitive that brain imaging reveals social relevance as another key factor driving people's decisions to share. At the intersection of self- and social relevance, we wonder: What will my choice to share say about me? What will others' reactions be? And, crucially, how will it affect our relationships?

I remember the first time I met Ruth Katz, a pioneering musicologist married to my late colleague Elihu Katz. We were at an event honoring Elihu's contributions to the field of communication, and Elihu was busy laughing and joking with the groups of people who had gathered around him. Although I didn't know her, Ruth welcomed me to sit next to her and chat on one of the curved cushioned benches of the Annenberg School's basement atrium. She looked out over the groups of students balancing plates of food and wine glasses and at Elihu with our colleagues beside the podium at the center of the room.

I figured we'd make small talk, since we didn't know each other well, but Ruth asked me about my life, looked directly at me, and offered advice. Kids, friends, and careers demand a lot of time and energy, she said, and sometimes people share their intellectual ideas only with work colleagues and don't discuss the ideas that drive their curiosity and career passions with their partner. She looked me in the eyes and told me that one key to a good marriage is making your partner your "primary audience." According to her, working with Elihu to make each the other's primary audience had enriched both their careers and their lives together.

Luckily for me, my partner, Brett, is a stellar "primary audience" when I want to talk about the brain or about things that are puzzling me at work. I also think back to Ruth's words when Brett wants to talk through his cryptography work in more technical detail than I signed up for (now we have a rule that he can go into as much detail as he wants about math, as long as he is playing with my hair). But I also think of them when it comes to the more seemingly mundane items Brett shares in our life together—these, too, are overtures and opportunities to connect.

For instance, Brett recently texted me a link to an article about the "hilarious and weird email sign-offs" of Gen Z workers. I learned that one such employee named Celine closes her professional correspondence with "Seeyas later," while another named Bryant uses "F*ck you, I'm out." Other sign-offs include "That's all" and "That's about it. Ummm . . . yeah." I figured Brett shared the article with me to highlight our shared experience of aging out of knowing what is cool. Sure enough, I rely on my much younger sister and Gen Z lab members to explain to me the often counterintuitive generational shifts in important matters like which emojis are acceptable to use (did you know that many young people read smiley faces as patronizing?).

Later that day, at home, it became clear that the article had struck a nerve with Brett, and he wanted to talk about it. "Why are you so interested in the Gen Z email sign-offs?" I asked. I had thought he was just trying to make me laugh by sending it, since we both get emails like this working with younger people. Instead, in a matter-of-fact tone, he explained that reading the compilation of sweet and silly email sign-offs made Gen Z feel less threatening to him. A rash of recent news articles detailing how members of Gen Z were increasingly prioritizing family, friends, and fun over long hours at work had made him feel like his own workaholic tendencies were being judged—like this younger generation was telling him that he had been duped, sold a lie about what it means to be happy and healthy. Yet, somehow, these email sign-offs that BoredPanda.com had compiled made him feel differently about his younger colleagues: less like they were secretly scorning his lifestyle choices and more like they were simply having fun at their jobs—telling it like it is. In turn, I shared that I love the ways that younger folks in my lab have pushed for more work-life balance; I think it is an important element to making the academy more equitable, and it makes me feel freer to relax too.

We saw earlier how sharing information about ourselves helps us strengthen our bonds to other people, but sharing other kinds of information—ideas, stories, news—can do the same. Over the course of our relationship, I've become more aware of the ways that Brett uses article sharing to try to connect with me. He appreciates when I read the full articles he shares and sometimes feels slighted when I admit that I haven't gotten the chance to open one yet. Thinking of Ruth— nudging me to not only treat Brett as a primary audience but also to be receptive and encouraging when he treats me as his—I get it. The articles are chances to connect with each other, to strengthen our bond and learn even more about each other. And according to my team's research, other people use sharing to deepen their connections too.

A team led by Joe Bayer, with Matt O'Donnell, Dave Hauser, Kinari Shah, and me, put the idea to the scientific test by inviting college student volunteers to play a game called Cyberball in our lab. Cyberball is a computer game of catch between three players, who throw a virtual ball back and forth to one another. The volunteers in our study believed they were playing with two other volunteers, but in reality the other players were controlled by a computer. These "players" were preprogrammed to play either a "fair" game of catch, where they throw the ball to everyone, including the human participant, equally; or an "unfair" game, in which they throw it to the participant a few times, but soon only throw to each other, leaving the real player out. This might not sound like that big of a deal, but people end up feeling left out, and it doesn't feel great.* We were thus able to simulate feelings of being included or excluded, kind of like how Brett felt when threatened by Gen Z's take on work.

After they were finished playing the game, we asked the volun-

---

* Don't worry—at the end of the study, we explained to the participants what really happened to help them feel better.

teers to help us with a task they thought was completely unrelated—beta testing a new news-sharing app. We prepopulated the app with a range of people whom the volunteers had named as close friends and family, as well as with friends or family they reported being less close to. We found that in this context people generally shared the most news articles with their close friends and family members, as you might expect—but after they had been excluded in Cyberball, they significantly increased the amount they shared specifically with close friends. In the face of a social threat (even something as trivial as being left out in a computer game with strangers), our volunteers sought to share, perhaps as a way of connecting with their friends, reinforcing their bonds, and feeling better.

It's therefore natural that regions of the brain that help us understand what others might think and feel—that is, the social relevance system—also often track people's interest in sharing ideas. Think about some of the things you've recently shared or that others have shared with you. Laura recently shared a podcast episode with me about parenting, for instance. She knew that the podcast—which discussed how we can parent in a way that gives kids respect, even when we don't always get it in return—was relevant to my daily life as a parent of young twins. How should we think about setting boundaries? How should we behave when we can't control other people's behavior? I found the episode thoughtful and funny, and although it didn't offer any truly groundbreaking advice, I thought it was entertaining and felt it was relevant enough to my parenting experiences to share with others as well. I decided to share the podcast with Anna and Ashley, who both have kids the same age as my twins. I didn't share it with Emma, who is kid-free and who I therefore guessed wouldn't find it as interesting. When Ashley told me that she had seen our hometown in a new HBO show, though, I immediately forwarded the screenshot to Emma, knowing she'd be psyched to see our old haunts on prestige television.

We all make such calculations when deciding whether to share something, and with whom. For the most part, these decisions don't require deep deliberation; I automatically (and only semiconsciously) made guesses about the social relevance of the podcast and the HBO show in deciding whether and with which of my friends to share them. Will Anna like this? Will the show make her laugh? Will the parenting advice seem obvious? What will it make her think about me and about our relationship? Even if we aren't aware of it all the time, we can see this process play out in the brain in the activation of the social relevance system when people are making sharing decisions. Interventions to change people's perceptions of a message's social relevance (for example, asking people to write a post that would "help somebody" rather than just describing what the article is about) also increase activation in the brain's social relevance system (as well as the value and self-relevance systems) and increase people's motivation to share.

Thinking about social relevance can also go beyond a desire to help and bond with others. It can also include wanting to gain status, look cool or smart or capable, or persuade others. Indeed, in the same research in which we learned that self-relevance can increase people's interest in sharing, we also tested social relevance's broad effects. We found that just as simple prompts that invite people to include content about themselves can encourage sharing, so can similar invitations to customize content for their network. This could entail highlighting anything from ways the content might make others feel ("whose day can you brighten with this news?") to what their friends stand to gain from the information ("tag a friend who needs to know this, and say why"). In addition, paralleling the effects of self-relevance, when people rated the content as more relevant to people in their networks, they also showed more activation in the brain's social relevance system.

But what if you have the option of sharing anonymously? Would the social relevance system behave differently? In a study out of Shenzhen University, researchers Fang Cui, Yijia Zhong, Chenghu Feng, and Xiaozhe Peng showed students news stories while their brains were scanned. Some of the stories reported on moral acts (people helping, saving, or donating to others), and some reported on immoral acts (people hurting, abandoning, or cheating others). The students had the opportunity to share these stories with others, but there was a twist: for half of the news stories, the students were told that the story would be posted anonymously, but for the other half, they were told the story would be posted along with their real name.

The researchers knew that self- and social relevance are important motivations for sharing, but they wondered whether the ability to post anonymously would change people's tendency to engage the social relevance system. Put simply, does anonymity make us less likely to use the brain systems that help us understand what others think and feel? They found that, overall, the students shared the moral headlines more than the immoral ones. This was particularly true when they posted using their real name, and was also reflected in the activation of their social relevance systems. Then, the team used a noninvasive brain stimulation technique to change activation of the social relevance system—and people responded to this activation accordingly. When brain stimulation dampened activation within a key part of the social relevance system, people showed less concern about posting immoral stories with their real name. In other words, when the research team disrupted people's social relevance system, it seemed to make them less cautious about what kinds of content they shared online.

When people think many others share their opinion, it also changes the way they use their social relevance system. My former graduate students Chris Cascio and Elisa Baek have run studies showing that people are more willing to share information about products like mobile game apps when they believe that others would also recommend the same thing. The same kind of psychology and neuroscience that we explored earlier in the book when I came to find Benedict Cumberbatch more attractive after learning the popular view can also alter preferences for what to share and recommend, along with the corresponding responses in the brain. People generally prefer to share ideas that are already somewhat popular.

This tendency can backfire when our beliefs about what is popular are mistaken. Research on pressing societal issues such as climate change, for example, has shown that people often incorrectly believe that although they themselves care about global warming, others do not. This phenomenon, which scholars call *pluralistic ignorance*, makes people less willing to speak up and share their views on topics like climate-related policy because they underestimate how much others will agree with or value their opinion (think back to the story of the emperor's new clothes!).

But other studies have shown that highlighting the number of people who have already shared or taken specific actions, such as adopting pro-environmental behaviors, can sometimes increase our motivation to do the same. Research led by Stanford psychologists Gregg Sparkman and Gregory Walton shows that in the early stages of building momentum for an idea or cause, highlighting the rising trend (more and more people are coming on board) increases people's willingness to act. Helping people feel part of a bigger whole can also make them not only more likely to take action personally but also to share with others and build even more momentum. Seeing yourself as part of a growing social group that cares about the environment, for example, is associated with higher pro-environmental intentions and behaviors. Therefore, just as simple prompts that invite people to include content about themselves can encourage sharing, similar invitations to join a growing group can motivate people to take action (for example, join a growing group of people who are contacting their representatives to demand change).

## BROADENING THE BEAM

I was sitting on the couch in my office when my then-doctoral student Christin Scholz stopped by to talk about the study we were planning. Beyond individual decisions to share, I wondered what makes some ideas likely to be shared by bigger groups of people, while others fall flat. I also wondered if it was possible to scale up the findings from the TV show study I had run as a graduate student and if there were commonalities in the brains of diverse audiences that made some content more appealing to share—not just in the lab environment but in real life. Would the same principles work when ideas travel from person to person in the real world? Would activity in an initial sharer's brain predict the sharing decisions of a series of subsequent people, as the interns' had for the producers in my initial study?

I had been thinking about investigating what makes people share through the lens of mobile game apps, but Christin had another idea that she argued could have more impact on people's well-being. She suggested that we look at what happens in people's brains when they read health news headlines. News often changes how we think about the world around us, and there are entire sections of the newspaper devoted to people's health—think of this as a daily health intervention, she said. Luckily, at the time, the *New York Times* made it possible to access statistics about each of their articles: how many people were reading each one and how many were sharing them. Bingo—an objective measure of large-scale sharing.

And so our team set about exploring whether it would be possible to forecast the sharing of health news articles around the world based on the brain responses they elicited from only a small number of people. We recruited two groups of about forty people each in Philadelphia and scanned the participants' brains while they read article headlines and rated how likely they would be to share them. We used real articles that had been shared to differing degrees: some articles had only been shared a few dozen times, whereas others had been shared thousands of times. To see if the people whose brains we had scanned picked up on the differences between the articles that had been shared a lot versus those that had not, we averaged data from their brain responses to arrive at a number for each article, representing how strongly the volunteers' value, self-relevance, and social relevance systems had responded to the headline. Finally, we compared these results with the statistics for how often those articles were downloaded and shared via email and social media among the global population of *New York Times* readers.

Remarkably, we found that brain responses from these two small groups in one American city helped us predict which stories were most likely to be shared across the world; articles that elicited the most brain activation in our small group of volunteers also tended to have the highest share counts online. Despite the seemingly large differences and variability in what people care about, people's brains seemed to converge on what would be popular.

Follow-up analyses of the same data showed that within these small groups, some people's brains were more predictive of wider large-scale sharing than others'. When we first set up the study, some of our research team advocated that we limit our recruitment of volunteers to frequent *New York Times* readers. The logic was: why waste resources scanning the brains of people who don't care about reading the news? For various logistical reasons, we decided to let anyone who could safely enter the scanner and wanted to participate be part of the study. I'm glad we did. When my then-postdoctoral fellow Bruce Doré dug into the data, he discovered that our volunteers who were frequent *New York Times* readers showed high activation in the value system to almost all the articles we showed them. Of course they did: they loved the *New York Times*! But this also meant that their brains weren't ultimately very good at distinguishing between viral hits and the articles that only the most committed readers love. Conversely, the brains of people who infrequently read the *New York Times* were more discriminating. When we saw greater activation in their value systems, that typically meant that the article was more likely to be shared widely around the world. In hindsight, this made sense: for something to be widely adopted, it needs to be liked not only by the committed people who are already disposed to love it but also by those who are less involved in that domain.

We can see how corporations have tapped into this to expand the market for their products. Consider the way Apple revolutionized consumer electronics by coming up with user-friendly interfaces and capturing the imaginations of not only tech enthusiasts but also broader audiences, or the way that Fitbit popularized step counting and brought focus on activity for people who weren't athletes. In considering how to address big societal problems like climate change, we would therefore do well to design products and technologies that are practical and cost-effective for people who aren't particularly eco-conscious and who might prioritize other factors in their decision-making. In related messaging about how and why to take action around climate change, it might also be useful to look at value system activity in people who are not predisposed toward eco-consciousness in order to predict how widely a particular intervention might be adopted.

In the years since our early work on the neuroscience of information sharing, we have run several more studies to explore how widely the value calculations of one group might represent others. Christin Scholz is now a professor of persuasive communication at the University of Amsterdam, where she runs a lab like mine. Although she sometimes reads the *New York Times* there, other news outlets carry more weight in Europe. Our original study captured the sharing of *New York Times* articles around the world, but Dutch people naturally don't make up the core of *New York Times* readers. If we ran a similar study including people from the Netherlands, we wondered, would their brains' value, self-relevance, and social relevance responses still track the articles' success like US volunteers' had? In other words, would our prior research findings translate to another culture, like the Netherlands, that has some commonalities and some differences with the United States?

Similar to the original study, we scanned people's brains while they read article headlines and asked them to rate how inclined they were to read the article themselves, and then looked at whether the brain activity we recorded and self-reported ratings predicted how much each article had really been shared on Facebook.

First, we found that despite reservations we had had about how the pandemic (it was 2021) and other recent events might influence people's states of mind and thus what we found, our previous discovery held: the collective brain activity of our relatively small group of volunteers was still a bellwether for how widely articles about health and climate change were shared on a much larger scale, among thousands of people. But when we compared how well the brain signals and self-reported ratings of both groups predicted virality, things got really interesting. Although the Americans' ratings tracked with how widely the articles were shared online, the Dutch participants' ratings did not—that is, a Dutch participant might say they were very interested in reading an article that was not shared very widely in the population of *New York Times* readers, or they might scoff at an article that had been widely popular. But the Dutch participants' brain signals *did* correspond with how widely the articles were shared worldwide. This meant that the Dutch volunteers were thinking about the articles in ways that made their conscious self-reported ratings less predictive of what articles would be shared, but their brains were still providing useful information.

To me, this suggests something intriguing about the underlying nature of what we value and how we decide to share. Even though Dutch and American people showed differences in what they told us they wanted to read, the responses we recorded in their value systems were more similar. Why might this be?

We don't know yet, but I'm excited to run more studies to find out if the brain responses are tapping into more fundamental needs that we all have regardless of our conscious thoughts about the content we're sharing (like the prospect of bonding or looking good) or if there are other sources of value that are shared more universally across these two cultures when people decide to share. I find this possibility intriguing—that perhaps the ideas that tap into activating our underlying self, social, and value systems are more similar than what we present on the surface.

## THE POWER TO SPREAD LIGHT

Understanding how self- and social relevance motivate people to share information with others might help guide us in offering people our full attention, like Theo did when he asked how my day was or like Eric did in planning his Fast Friends date with Laura. Understanding these motives can also help us make sense of our urges to turn to our phones or social media, even when we might otherwise want to be present in the moment.

When Eric proposed to Laura a little less than three years after sending the virtual rose, I was the first person in our family to find out. As we approached my house, where I knew they were about to tell my grandmother and mom, I got my phone camera ready. I knew that my mom and grandma would want to email everyone they knew, and after the initial excitement settled down, at the Thanksgiving table, despite our family's "no phones at the table" rule, I couldn't help texting photos to all my friends as well (my family graciously gave me a pass here, but I probably could have waited the extra half hour to do it after the meal).

But as many of us know, the kind of connection we seek through reaching out to others and sharing isn't only about the joyous times, the falling-in-love times. It also comes to the fore during the hardest times we face, during times when we want to create community and create change, as became clear when my dear friend Emile was diagnosed with glioblastoma, an aggressive form of brain cancer.

Less than two weeks after the diagnosis, Emile gathered his closest collaborators in his living room for a chance to share. We were neuroscientists, political scientists, leaders of NGOs working in conflict regions around the globe—those in Philadelphia sat shoulder to shoulder and others Zoomed in. Emile was energized and focused, his eyes glimmered, and he had a fresh scar that began near the nape of his neck and extended about five inches up his skull. I could still see the staples.

Emile was my colleague. I had hired him to work in my lab four years before, and though he'd moved on to open his own lab, Penn's Peace and Conflict Neuroscience Lab, we'd been collaborators ever since. He said he wanted to focus the meeting on his vision: promoting peace in Colombia, reducing dehumanization of people in conflict or who had been historically marginalized, taking on increasing political polarization in the United States and beyond. This meeting was one of the first things he organized postsurgery, because it was that important to him. He knew the only way forward on these projects was to share them with everyone gathered.

He had made a slide presentation, and he brought us up to speed on how he was thinking about his life's work. He understood that the situation was a strange one. The people he'd gathered were, on one level, grieving—he likely had less than two years to live. He understood that kind of grief and the value of being able to share it communally. "This is sacred and cherished and such a beautiful part of humanity," he said. "I don't want to take that away from you. But I want to give you some of the perspective that I have." He went on to talk about what he hoped we would carry forward from his legacy: "Our goal should be more dramatic than just doing good science, although that's important and wonderful and good," he emphasized. "We have the potential to do more. We have the potential to walk through darkness and spread light."

To realize this potential, at this meeting, and in the months that followed, Emile made sure that we were connected not only to the ideas but to each other. Toward the end of the meeting in his living room, through tears, he explained how we might access the greater potential he envisioned to spread light. "The nice thing is that this force is in us and communal. It's not owned. And the best way to activate a communal force is to be a community. So that's why we're here."

When a choice is about *connection*, it resonates on a different level.

Many people have made the mistake of trying to get people to support a cause simply by sharing information about it, rather than offering people the chance to see themselves in the messaging or offering some social value in sharing. But Emile did something he had been doing his whole life, and that was part of what made his ideas so infectious: he brought us together, shared with us not just information but his connection to it, and impressed upon us that these ideas were how his work would live on. He made it about us, and helped us build the bridges to one another, so we could call on each other, in his absence, to advance the work and his broader goals.

We have the power to influence those around us with what (and *how*) we share and in the opportunities we provide for others to share with us. As we'll explore next, in doing so, we have the chance to illuminate new pathways of action that shape the culture we are part of.

# 9

# I Am the Beginning

MARIA RESSA TOOK A deep breath, clearly brimming with emotion as she stood at the podium to accept the 2021 Nobel Peace Prize. She was being lauded for her "efforts to safeguard freedom of expression, which is a precondition for democracy and lasting peace." In her speech at the lectern, she recounted the harrowing work of being a journalist for three and a half decades working in conflict zones and documenting disasters in Asia—both natural and political.

You might think that someone about to receive the Nobel Peace Prize must have led an extraordinary life with extraordinarily high-stakes decisions—and you would be right. After serving as CNN's bureau chief in Manila and as a lead investigative reporter in Asia, she cofounded Rappler, a news website based in her home country, the Philippines. Not afraid to hold government officials accountable, her team at Rappler had been critical of then-president Rodrigo Duterte's government. In 2019, she was arrested for cyberlibel (online defamation), a move seen by many around the world as politically motivated retribution for speaking out. So, yes, Maria *did* make extraordinary high-stakes choices. But she also credits the smaller, everyday choices that she made along the way for making her who she is. As she described in her autobiography, *How to Stand Up to a Dictator,* she chose to learn, to prioritize integrity, to be vulnerable and honest, to embrace fear, and to collaborate. These choices carried her though childhood experiences like moving from the Philippines to the United States, making friends, and working hard to learn English and excel at school. These same principles resonated throughout her career, as she earned a Fulbright that she used to travel back to the Philippines, where she began the work that landed her in Oslo.

When she first returned to the Philippines in 1986, an old friend rekindled their relationship and invited her to spend time at Peoples Television 4, a revolution-ravaged news station with broken lightbulbs and hallways that stank of cat pee. At the time, in that place, the project of the news itself was no small thing—the Philippines was fresh off a dictatorship that held tight control over the news media. So when Maria was offered a job there, you might imagine the constriction that she could have felt—walls pressing in—but she saw possibility.

Maria looked around at the buzz of people working together, scripts being written just minutes before airtime, run over to the anchors who would read them live on air—"the first page of history being created with tremendous impact." Maria said yes, and threw herself into the work of producing the news—the work of her career. It was work that she came to realize had been shaped, from the beginning, by the cultures she was part of.

With her Nobel Prize in hand, she said: "At the core of journalism is a code of honor. And mine is layered on different worlds—from how I grew up, when I learned what was right and wrong; from college, and the honor code I learned there; and my time as a reporter, and the code of standards and ethics I learned and helped write. Add to that the Filipino idea of *utang na loob*—or the debt from within—at its best, a system of paying it forward."

*Utang na loob* is an example of the type of cultural value that shapes our smaller, daily value calculations. A person in a culture that emphasizes the importance of a debt from within might feel a greater pull to repay that debt when faced with a choice between helping someone else or focusing on oneself, compared with someone from a culture that places less emphasis on relationships and focuses more on individual success. Maria's interpretation was that "you are responsible not only for yourself but also for the world around you, your area of influence." She knew that our cultures are made up of the beliefs, preferences, and behaviors of individuals within it—including yours and those of the people around you.

Indeed, there's a give and take between our cultures, the personal values we hold, and our day-to-day value calculations. On the one hand, culture shapes the norms and identities that influence our brains' value calculations, shifting the balance of what we pay attention to and value and informing the decisions we make. This includes everything from what we think is "normal" and tasty to eat for breakfast, to whether we think democracy is the best form of governance, to how we think it is appropriate to respond if someone pays us a compliment. On the other hand, cultural norms are not wholly deterministic. The decisions we make may conform to cultural norms or not; when others see what we do, it influences how *they* calculate value and behave—creating feedback loops that potentially shift cultural norms over time in *our* areas of influence.

We learned in the first parts of this book that observing peers influences our value calculations and more broadly what we pay attention to. Now we'll see what this all adds up to on the macro level: how cultural norms shape our value calculations and, importantly, how our individual choices have the power to shape—and reshape—cultural norms.

## CULTURE'S INFLUENCE

Think about your favorite comfort foods, how much autonomy you think young people should have in choosing what career to eventually pursue, or whether and whom to marry. Now think about how someone in another country, on the opposite side of the globe, or even someone who lives in a different part of your own country, might answer those same questions. There's a lot of variety in what people think is normal and desirable, and research suggests that people from a wide range of cultures soak up similar values to other people in that same culture. These cultural values and practices not only shape our thinking and behavior but also change how our brains work.

For instance, it's commonplace that, on average, people from Western cultures tend to prize independence—individuals pulling themselves up by their bootstraps, stars achieving results on their own—whereas people from East Asian cultures tend to emphasize interdependence—individuals prioritizing social relationships and the collective good, and groups of people achieving larger goals together. Research in the lab has shown that these differences are reflected in how people use their self- and social relevance systems. Across a range of tasks, people from East Asian cultures tend to show more activity in regions within the social relevance system, and people from Western cultures tend to show more activity in regions within the self-relevance system.

For example, when the researchers Yina Ma and Shihui Han, at Peking University, compared the brain responses of Danish and Chinese volunteers who were rating how much different words described them, all the volunteers showed activity within the self-relevance system, as you would expect. Yet the Danish volunteers showed *more* activity in core parts of the self-relevance system, like the medial prefrontal cortex. Conversely, when the volunteers were asked to think about their relationships with others—making judgments about whether terms that link people to one another, like "tenant" and "professor," described them—the Chinese volunteers recruited parts of the social relevance system, including the temporoparietal junction, more strongly than the Danes. In a survey, the Chinese volunteers on average placed more value on interdependence than the Danish volunteers did. The underlying neural processes that shape how we think about who we are are themselves shaped by our cultures.

Culture also informs the value calculation. Neuroscientist Eva Telzer and her colleagues, then at UCLA, found that young adults from Latinx and European American backgrounds processed choices to donate money to their families differently. Compared with European Americans, on average, the Latinx young adults in the study reported feeling that their family formed a greater part of their identity and also reported helping their families more with things like household chores, sibling care, and business assistance (for example, helping someone in their family with government forms or translating text). To uncover how these cultural differences relate to people's daily value calculations, Eva's team scanned all the volunteers' brains while they made decisions between contributing money to their families (some of which required them to give up their own money) or earning money for themselves. Though the two groups tended to contribute similar amounts to their families in the lab, the scientists observed that the young people's brains processed the choice differently. The Latinx volunteers showed greater responses in the brain's value system when giving up their own money to help their family, whereas on average, the European Americans' value systems responded more to personal gains. This suggests that—despite making similar actual decisions in the lab—the different groups valued the choices differently.

The effects also went beyond the lab. For both groups, the more activation they showed in the value system when making costly donations to their family, the more they tended to help their families day-to-day outside the lab. The young people who said that their identities were more connected to their families showed the strongest effects. This suggests that culture shapes not only expectations about what people will do—in this case, help family—but also how meaningful and rewarding they find it.

Together, these studies and more highlight that culture influences the way our brains process self-relevance, social relevance, and value and, in particular, when each of these processes might be given more weight. Think back to Jenny Radcliffe breaking into the bank in

Germany: she focused on making a scene in part because she knew that German culture tends to prioritize public order. Jenny (correctly) predicted that drawing people's attention in the lobby would embarrass the German guard and motivate him to let her in. The same approach might not work in a Mediterranean country, where people are generally more relaxed and less concerned about a chaotic lobby. Knowing the cultural context was key to understanding the guard's likely value calculation.

Extending our earlier insight that we sometimes think we know more about what others think and feel than we actually do, these findings highlight the need to take extra care in working to understand the background assumptions and decision-making processes of people who have different cultural backgrounds. How do your coworkers like to receive feedback? How appropriate is it to express a strong opinion at the dinner table when you first meet your new partner's parents? How does your friend feel if you strike up a conversation with them in the library or a coffee shop, where others are reading nearby? Rather than assuming others would answer these questions the same way you do, it is helpful to gather data by observing and asking.

### BREAKING THE MOLD

As powerful as the influence of culture is on us, it's not deterministic. We have our own complex identities within our cultures—constellations of influences that may move with or against the grain of the larger culture. Age, race, gender, sexual orientation, (dis)ability, education, career stage, and other aspects of our identities shape what others expect us to do, how we are rewarded or punished for our behaviors, and also what we expect of ourselves. Different contexts make different identities more salient and give us opportunities (or limitations) to focus our attention on different possible choices and priorities.

Research shows that people's preferences can be directed based on what identity they focus on. Leor Hackel and Jay Van Bavel, at NYU, found that when Canadians answered questions that made them think about their Canadian identity, they found maple syrup— a classically Canadian treat—tastier than honey. But when they were primed to think about other parts of their personal identity (without focusing on being Canadian), they experienced maple syrup and honey as equally tasty. Cultural identity influences our value calculations, even for something as basic as what belly-timber we find tasty in a given moment. But even more important, this study highlights that our identities are multifaceted and that the context we're in can shape which aspects get the most weight—Canadians didn't *always* prefer maple syrup. Who we are and the social context we're in interact to shape what we care about at a given time. If parts of our identities are responsive to social contexts, then how might these different identities and different contexts shape how our brains calculate self-relevance, social relevance, and value?

In research led by communication neuroscientists Arina Tveleneva and Chris Cascio, my team explored how stereotypes about gender interact with the cultural values of independence and interdependence when people make decisions. In this study, participants looked at a selection of mobile game apps and their descriptions and made initial decisions about whether or not they would recommend the game. Then they went into a brain scanner, where they were reminded of each game and their initial rating, but also told how their peers rated the game compared with them (higher, lower, the same, or no rating information available). They were then asked for a final rating. We wondered if gender identity, as well as stereotyped ideas about that identity, might affect people's behaviors. Specifically, would men who fit the stereotype of being more independent and individualistic, and women who fit the stereotype of being more interdependent and communally oriented, use their brains differently than people who defied these stereotypes when processing peer feedback?

We found that they did. Stereotypically congruent women (who were more interdependent) and men (who were more independent) tended to show more activation in the brain's social relevance system when they conformed to peer opinions of the mobile game apps. In other words, people who align with societally prescribed ways of acting have higher activation when they are doing what the group expects them to do. It might be easier for them to stay in alignment with group expectations, or they might view conforming as the most socially relevant path.

Conversely, independent women and interdependently oriented men showed the opposite pattern. They showed increases in the social relevance system when they were defying the group, but not when they conformed. It might be that the way they process social relevance didn't value conformity as much as going against the grain; defying mainstream expectations might feel more socially relevant.

Because of the dominant norms in the United States, although we may expect women to conform more and men to conform less on average, it was the interaction between the cultural expectations of the volunteers' gender groups *and their own positioning in relation to those expectations* that shaped how their brains calculated social relevance. When we bump up against cultural expectations, our individual identities might shape how we respond. It is the interplay of cultural influence and individual identities—not one or the other—that shapes how we respond to the world. In this interplay, there's great potential. The individual's incongruence with the whole can move the needle on larger cultural norms.


## OUR FRIENDS' FRIENDS' FRIENDS

When Maria Ressa decided to continue her reporting to stand up for the values she had learned from her family, community, and education and to pay forward the debt from within, she knew she was putting herself at risk. But she also knew that there were eyes on her—that her decisions would impact others, who would make their own decisions about the importance of freedom of expression, democracy, and access to facts.

She could have decided not to report on President Duterte. She could have stayed away from politically charged stories about drugs and violence. When the government began investigating her company in retaliation, she could have stepped back from investigative reporting altogether.

The stakes of our day-to-day decisions are probably less dramatic than Maria's—we may not face arrest for sharing facts or our opinions online—but what we share and the way we show up in the world can illuminate possibilities and paths that influence others, even if our light only begins by streaking through the smallest crack. That small glimmer of light can spread across groups of people. Cultural norms can change.

In 2016, a team at Princeton led by the MacArthur "Genius" Award winner and psychologist Betsy Paluck worked with kids in New Jersey schools to highlight the power they can have to affect the people and culture around them. Betsy's team worked with fifty-six middle schools across New Jersey that together enrolled over twenty-four thousand students. In some of these schools, the researchers randomly selected small groups of students to become what they called "seed students." These students worked intensively with the research team to identify and implement ways to take a public stand against injustice—in other words, they thought about how they wanted their school to be and how to best share that with peers. The students created online hashtags and paper posters to display around school and shared the hashtag slogans and photos of the students who came up with the ideas. The teams of students also gave out bracelets when other students were supportive of their peers or did things that discouraged harassment, bullying, violence, and so on, as a way of visibly recognizing and encouraging kindness norms.

All these efforts aimed at highlighting the social relevance of being considerate and standing up against harmful, antagonistic behaviors. The results were impressive. In the schools where seed students modeled the desired norms, disciplinary reports of student conflict declined by nearly a third over a year. Students in these schools were more likely to talk with friends about how to reduce potentially harmful kinds of conflict in their school, especially if they were direct friends with seed students. Students who interacted with the seed students also dramatically updated their perceptions of how much others disapproved of bad behavior, moving from believing that only "a few" students disapproved of racial and ethnic jokes, for instance, to believing that "about 75%" of students disapproved of them. Overall, the schools that received the intervention saw reduced average levels of disciplinary reports of peer conflict compared with those that didn't, indicating that when students think their peers disapprove of hateful behavior, they are less likely to engage in it themselves. Not only can individuals contribute to culture shifts; this study offers a model for giving people the tools to channel their power to shape the cultures they are part of. We have the power to shape local norms through our actions—and by helping others notice and use their power there as well.

We each have a role to play in shifting norms and changing people's expectations of what good (or bad) things might come from working together, even if in small ways and even if it is one step at a time.* This is particularly vivid when we think about children and how their values are shaped. For instance, the books we choose to read to our

---

* Having said that, the consequences for violating societal norms are not the same for everyone. Maria risked imprisonment and violence for her choices. To differing degrees, many of us are entangled in cultural contexts where going against the grain can come at high personal and social cost, and this is more pronounced for social groups that have less power and fewer resources. Building on what we explored about leadership and power earlier, when we have relatively more power, it can be invisible to us, and we may not think about how our habits and actions influence others around us. Yet these are the times when it is most important to listen, read, and find out what might support others in our communities.

kids, the movies they watch, and the toys they are given—all this shapes their ideas about gender, power, and other important concepts. Do men tend to be sweet caregivers? Do girls tend to operate power tools or skip school to go fishing? Who tend to be leaders and in what ways? The conversations we have with our children also help them understand what *we* think of these messages. Is it cool that the dad in "Rumpelstiltskin" sends his daughter off to work for the king spinning straw into gold in exchange for the opportunity to marry him, or would she have been better off avoiding a guy who is primarily interested in her for the money? Should she have been given a say in it?

Moreover, as we've learned, we are capable of focusing our attention on different parts of our environments. For this reason, calling attention to behaviors that are aligned with values we care about ("Did you see how that kid just shared his toy?"; "I'm happy to give you a bite of my ice cream, since in our family we share food") and talking about our decision-making process ("My friend was feeling stressed at work today, so I decided that other deadlines could wait and we went for a walk. It wasn't an easy decision, but it is important to me to show other people that I care about them") can highlight the values we want to impart to our kids. When we act in line with our goals and values, it can motivate others to follow suit. And when we tell them how we feel about their actions, it similarly changes how they see the social relevance of those actions.

The same basic premise works with adults too: our behavior can influence our friends and others around us. This happens with the small stuff; for example, if we share our love of a particular style of art with a friend, it is likely to change their underlying assessment of beauty, and learning what foods we love likewise increases how much others around us want to eat those foods. But even beyond directly sharing our likes, dislikes, and ideas, people around us also *observe* what we're doing—and this influences their actions.

Research led by Erik Nook and Jamil Zaki, at Stanford University, highlights two important facets of our potential influence on others. First, conformity is rewarding and valuable, even when there are reasons we might have personal preferences that differ from those of others. Imagine that you are at a restaurant with a friend, and you are each considering what to order. "I love the butternut squash soup here," you muse. You order the soup. "Great, I'll get the soup, too," your friend tells the waiter. What is going on in your friend's brain? Why would they choose to order the soup when you know they love pizza (also on the menu)? When hungry people in Erik and Jamil's study learned that their food preferences matched other people's food preferences, it increased activity in a key part of the value system that responds particularly strongly to rewards—the ventral striatum. This indicates that we sometimes assign more value to agreement than to our initial taste (matching your butternut squash soup versus getting the pizza). Further, the more they showed this reward response to being in line with the group, the more likely they were to change their food preferences when they learned that other people liked certain foods more or less than they did. In other words, the people whose brains placed the most value on being aligned with the group were most susceptible to peer influence.

Second, Erik and Jamil learned that if you look at a part of the value system that integrates different aspects of a choice (the ventro-medial prefrontal cortex), you can see the potential impact of peer influence on a person's behavior. In their study, the hungry volunteers initially showed more activation in the ventromedial prefrontal cortex when they looked at pictures of unhealthy foods like chips and candy, compared with healthy foods like fruits and veggies. In other words, as in research we explored earlier, their brains were initially more excited about the immediate reward of junk food. But then Erik and Jamil shifted the participants' focus by telling them about other people's healthier preferences. When they did this, they found that the volunteers' ventromedial prefrontal cortex then tracked their peers' opinions.

It also seemed to change their preferences in the longer term—the volunteers in the study later said they liked the foods that other people also liked more, and said they liked the unpopular foods less than they initially had, even though all the feedback was actually randomly generated by a computer (and therefore had nothing to do with the tastiness of the different foods). In the restaurant example, your preference for the soup changed your friend's underlying value calculation and their estimate of how much they might want squash soup. Obviously, this doesn't mean they won't ever eat pizza again, but when we hang out with people who tend to eat more (or less) healthy foods, it shapes our diet over the long term. Thus, the choices you make influence the choices that others make, and their choices influence even more people.

This extends far beyond food preferences; research also shows that larger moral norms are influenced by those around us. Taking your kids with you if you volunteer in your community helps them see neighbors and friends working together and increases the social relevance of doing the same. Wearing your "I voted" sticker, as well as giving the people you manage time off to vote, signals the social relevance of participating in our democracy. As we explored in the previous chapter, sharing your interest in a particular cause online can increase others' perceptions of social relevance. Consciously and unconsciously, your actions change how valuable it is for others to act in certain ways.

Even simple actions like showing up and providing "likes" or appreciation influence people's value calculations. In a study at Leiden University, neuroscientists Jorien Van Hoorn and Eveline Crone sought to learn if teenagers would make more generous donations when their peers were in the room. The teenagers played a game where they could choose whether and how much to contribute to a pool of resources that benefited everyone. Indeed, being observed changed the volunteers' behavior. When they believed they were being watched by their peers, they showed more activation in self- and social relevance brain regions than when they played with fully anonymous peers, and the more they showed this brain activation, the more they donated. And it seemed this giving could be even further encouraged. When the peers provided feedback, in the form of "likes" for larger donations, the teens showed even stronger brain responses in the social relevance system and were even more likely to increase their donations to the common pool. In other words, giving people feedback about attitudes and behaviors that you believe should be valued shapes the way their brains calculate value.

Other motives can hijack the same brain systems to different ends. Consider the ways that social media algorithms often amplify outrage by promoting cultural norms that condone it. Research led by William Brady and Molly Crockett at Yale showed that across thousands of Twitter users and millions of tweets, when people got positive feedback for expressing outrage, they were more likely to display it in future tweets. Just as people behave more generously when peers "like" their donations to a group cause, people also amplify their outrage when platforms incentivize it. Getting rewarded for expressing outrage might make you express more over time, feeding the norm that it is appropriate to do so and creating a reinforcing feedback loop. In turn, the more time you spend on social media reading political news, the more you overestimate the amount of outrage others are feeling, and the more you think you should be feeling it too. Liking and sharing others' outrage teaches not only you but also others to amplify similar expressions of emotion.

Value calculations are transmitted from brain to brain when we share ideas with others, meaning that we cocreate norms with others; in Maria's words: "Everything we say or do impacts our friends, our friends' friends, and even our friends' friends' friends." As we've seen, communication can create synchrony between people's brains. This means that when we actively tell other people about our preferences, some of the same patterns of brain activity that produce those preferences in our value systems get reproduced in their minds as well. When we watch Maria Ressa at the podium accepting her Nobel, our brains replicate some of what happened in her brain as she stood there and also align briefly with other audience members watching the same speech.

To some extent, Maria knew this—she had used the power of social influence before, in the May 2010 presidential election in the Philippines. For her team's Get Out the Vote campaign, she took advantage of the power of social influence. She set up a crowdsourced citizen journalism program on politics and social concerns and used a simple slogan—"Ako ang Simula"—which literally means "I am the beginning" and in spirit means "Change begins with me." It was inspired by an idea often credited to Mahatma Gandhi—"Be the change you want to see"—and even further back to Plutarch: "What we achieve inwardly will change outward reality."

The research bears out this message. Individuals pick up the norms of the environment they are in, but can also spread those norms to new environments. In one study, in which volunteers were randomized to play a series of games in an environment built to support cooperation or competition, the folks who played with other cooperative players behaved more cooperatively themselves, while the folks who played with more competitive players behaved less cooperatively. When the volunteers later played games in a new environment, the people who had been in the cooperative environment were more likely to behave in cooperative, prosocial ways, highlighting one way that norms might spread from one context to others.

If you are collaborative with your team at work and encourage others to cooperate by praising their efforts to work together, they may also be more cooperative when they work with others on different teams—even if the norms there are more competitive. But if you behave competitively or incentivize a more competitive atmosphere, that, too, might spill over to other teams within your organization. The choices we make individually influence others, and their choices influence us. As more and more people adopt particular ideas and behaviors, culture shifts—and this changes the way that our social, self, and value systems take in information in the first place.

This played out in Maria's campaign as well. After four months of organizing lectures, talks, youth activist speeches, concerts, and workshops, the Commission on Elections in the Philippines asked them to slow down—"Its systems couldn't keep up with the number of voter registration applications pouring in." This was a real change in civic engagement, and it began with Maria's small team.

The ways that people wield power can result in harm, as we discussed earlier, but it can also result in increased well-being (as with the kids in New Jersey) and increased civic engagement (as with Maria Ressa's efforts). Knowing that what we express can be contagious, we can take steps to be more aware of which sorts of behaviors and rhetoric we reinforce and which we discourage. Doing so may lead those around us to be more aligned with the values we want to see in the world.

## MAKING SPACE, TAKING SPACE

Sometimes we have the time, the energy, the luxury of making our decisions align with a bigger-picture set of goals and values; sometimes we hurry along, attending to whatever inputs to the value calculation happen to demand our attention. We have dinner to get on the table, Legos on the floor, friends texting, and deadlines to meet. Some days we are just doing what we can to inch slowly forward.

But when we have the time and space to do so, we can also take a step back and reflect on how we spend our time and energy, whom we give our attention to, and how we might align those decisions with our bigger-picture goals and values.

As Annie Dillard reminds us in *The Writing Life*, the ways we spend our minutes, our hours, and our days add up to how we spend our lives. Would you rather go to the school play or the board meeting? Would you rather speak out when a colleague makes a mean remark or stay quiet? Would you rather work late or meet up with a friend?

Would you rather call your senator to make your opinion known or let someone else decide? As we've seen throughout this book, each of these choices is shaped by the inputs to the value calculation that are front and center when we decide. How might your choices change if you imagined all the people who are influenced by seeing your example? What if you're not alone in that cave, with the walls pressing in? What if someone behind you sees where your beam of light points? What if they follow?

Young Maria made the choice to take a job at a news station that had suffered under years of censorship, focusing on the possibility of the work. As the events of her life unfolded, she called upon the lessons learned during small moments in her childhood, the books she read in college, her early newsroom training, the practice of making decisions under pressure in the field, again and again. Toward the end of her Nobel speech, Maria looked out at the audience. "I didn't know if I was going to be here today. Every day, I live with the real threat of spending the rest of my life in jail just because I'm a journalist. When I go home, I have no idea what the future holds, but it's worth the risk."

When we make choices, others see them. When we see Maria Ressa speak out, it highlights the social relevance of standing up for democracy. When the Nobel Committee awards her the Peace Prize, it calls attention to, and endorses, her way of thinking and acting. It doesn't erase the systems she's fighting against or the structural constraints that can throw barriers in our paths. But if enough of us make a different kind of choice than broader systems advocate for, then we can shape the kinds of cultural norms that challenge them. Maybe, slowly, it could even lead us out of the cave. Slowly we realize that the walls don't have to be so tight, that we aren't alone, that each one of us has so many sources of light, illuminating so many possibilities. It's not only that we make ourselves with the choices we make—the choices we make, together, in the long run, create the world we live in.

# EPILOGUE

LONG BEFORE MY FRIEND and colleague Emile Bruneau gathered his colleagues in his living room to pass on his work, he already had a clear vision of how neuroscience and psychology could be leveraged to help people make different choices, to change and connect.* After learning about his brain tumor, I couldn't help but admire how, while staying rooted in his core values, he quickly pivoted his understanding of himself and his future to his new reality: one where his sight was fading and he didn't have long to be in this body. I watched him work closely with his family and collaborators to make so many high-stakes choices: What treatment options would they pursue in these last months of life? How would his children feel his presence after he was gone? How would his work on empathy and peace continue and thrive?

After Emile's diagnosis, I thought a lot about his brain—not just the tumor, but also what might be happening as he made all these choices. He was able to balance a focus on the present and optimism about the future; he possessed a core sense of self and an openness

---

* This clarity of purpose, which was beautiful, also made it easy for him to say no when people tried to pull his attention away from what he thought mattered most, and yes to the important things. He would drop everything to travel across the world to try to facilitate peace between groups in conflict. He would also drop everything to look at a bird that his wife had spotted out their third-floor window.

to change; a sense of how powerful it is that our brains allow us to imagine what others might think and feel and the awareness that this can mislead us; a talent for finding joy in sync with a wide range of others, without getting pulled into an echo chamber. I started taking videos of our conversations and spending as much time as I could talking with him about everything from parenting to peacebuilding. You can study something in the lab for years and still not really witness the complex and disparate pieces of it come together in real life in this way. He made choice after choice, laser-focused on what was important to him. He made it look almost easy, though of course, for most of us, it is not.

It wasn't easy for me, a few years later, when Bev told me that we weren't spending enough quality time together. At that small request, I felt constricted and conflicted. I felt short on time making dinner for my kids and thinking about work deadlines and wanting to walk with my grandma. Yet noticing, growing, deepening our connections with other people is fundamental to our well-being, to our ability to innovate, to our sense of who we are and our ability to make choices that feel "right."

When Emile faced an even more acute limitation on his time, he responded so differently. He lived constantly in the expansion of possibility—leaning into his connections with other people and letting go of the strict boundary that many of us place on our presence, openness, love, empathy, and willingness to see ourselves as one with others.

One night, after my kids went to bed, I sat with Emile in his hospital room, playing my dad's guitar. My dad had recently died, and I was sharing my grief with Emile, who in turn was wondering about what life would be like for his kids after he died. Then, after a pause, Emile turned his head toward me and described how meaningful it had been to him, in the years after his mother's death, to continue to deepen his relationship with her.

At first, I didn't understand. "What do you mean?" I asked. "How can you deepen your relationship with someone who is gone?"

"The same way as you do with someone who is alive," he said.

Emile was always doing things like this—taking a situation that seemed fixed, bounded, sometimes impossible, and finding a new possibility inside it.

Emile reflected on what the psychology and neuroscience research tells us—that so many of the ways we engage with each other actually come down to engagement with the ideas we each carry, the stories and the practices, what we think and what we think someone else might be thinking. We don't spend that much time physically with most people even when they are alive;* our relationships with them play out mostly in our minds. Yet these relationships influence how we understand ourselves today and how we make the choices that determine who we will be tomorrow.

It seemed that Emile's understanding of the brain was illuminating, comforting, maybe in some cases even empowering to him, as I often find it to be. I hope it was, and I hope that this book might be for you. Governments and research teams are currently trying to figure out how to encourage and create more transparency in the calculations of artificial intelligence, hoping that we can better align their outputs with our human values. Maybe understanding the brain can offer a beginning point to do something similar within ourselves: to grasp some of the elements that the self-relevance system synthesizes to tell us the story of who we are; to begin to understand the algorithm that the social relevance system uses to make sense of other people's minds; to know some of the ingredients feeding into our value calculations and sense the weights assigned to each. Maybe we can even rebalance those weights for ourselves or others. After all, this work is fundamentally about helping bring people's daily behaviors in line with their vision for the world they want to live in.

In one conversation, Emile reminded me that since our cells

---

* Although Emile brought this idea to the forefront for me, people around the world have connected and conversed with their ancestors in these ways for millennia, through personal practice and rituals designed to build these connections.

refresh and turn over so quickly, what makes us who we are isn't the physical combination of cells in our body but rather the *patterns* of how they work together. And certainly within the brain it is the patterns of firing and connection rather than any specific collection of neurons that render our thoughts and feelings and who we are. As a neuroscientist who studies communication and how ideas and behaviors spread, I'm also filled with wonder thinking not only about the ways that pieces of who we are are encoded in the patterns of firing in our own brains, but also about how those patterns get transmitted beyond. The patterns our thoughts, feelings, and actions set in motion ripple out in their impact, growing, adapting to new environments and challenges in this world. This is part of why I do what I do, part of why I wrote this book, and part of what I hope you will take from it.

In this way, even when we feel alone, we are still interconnected with others, and parts of us are distributed across people and time. Walking hand in hand with Bev under the canopy of trees near her house, she tells me that she wants us to have the party we're planning for her hundredth birthday, whether she's there or not. I tell her that she'll be there. We will celebrate life, this life in which we are inextricable from each other.

# ACKNOWLEDGMENTS

This is a book that is, in part, about how we imagine who we are and what is possible. It is also a book about the influence we have on one another in imagining those possibilities. I am deeply grateful to the many people who made this book possible and helped me imagine what could be. Thanks to my editor at W. W. Norton, Jessica Yao; my agents Celeste Fine and Jaidree Braddix at Park & Fine Literary and Media (PFLM); and Katie Booth, who provided tremendously thoughtful edits and made the writing better. Working with the four of you individually and as a team has been an incredible gift. Your insightful feedback, edits, expert guidance, and incisive questions have not only shaped the book but also pushed me in how I think about the research I do. I love your unique and collective senses of humor and appreciate your values and skill in working collaboratively. Thank you for believing in this project and for all the ways you have provided instrumental and emotional support. Celeste and Jaidree, thank you for being the best advocates for the work, for all of your insights, and for your generosity with your time. Katie, thank you for making me feel seen throughout this process. In addition to being a brilliant writer, your expertise as a coach and ability to understand where I wanted to go and why have wowed me. Jessica, I have learned so much from you about how to write a book, and I'm always floored by your ability to put your finger on the right question. Thanks also to

Annabel Brazaitis and Yumiko Gonzalez Rios at Norton, John Maas at PFLM, and Louisa Dunnigan at Profile Books for helpful input, and to Janet Greenblatt for copyedits.

Ally Paul, Evan Wilkerson, Victor Gomes, Nikolas Silva, and Selam Bellete helped get the details right—thank you for fact-checking, reference checking, and creating resources that make it clear where the claims in this book come from, making the manuscript look good, and flagging places where I needed to rethink or rephrase. Thanks to Omaya Torres for creating the art in this book.

Molly Crockett, Sandra Gonzalez-Bailon, Rebecca Saxe, Christin Scholz, Dave Morse, and Alice Marwick were early readers and provided in-depth feedback on the manuscript, and the Annenberg Center for Collaborative Communication and Dean Sarah Banet-Weiser sponsored the book workshop that made it possible to get this critical feedback. I don't know what the book would be without this. Michael Delli Carpini, Uma Karmakar, and Adam Kleinbaum provided constructive feedback on early chapter drafts, and Dave Nussbaum shaped my thinking about communicating with broad audiences in many ways, including helping me think through several elements of this book. Dave, the service you have done in making it possible for scientists to share their work is incredible, and I'm so grateful that you are you.

Matt Lieberman introduced me to social neuroscience, guided me through graduate school and beyond, and modeled the importance of working as a team and sharing science with broader audiences. Matt and Elliot Berkman were excellent thought partners at the start of this journey. I'm so grateful to have gotten to work with you. Thanks to Bob Kaplan, Noah Goldstein, and Naomi Eisenberger for shaping my early thinking about research, and to Lauren Wong Bevand, Liana Epstein, Kathryn Brooks, Tristen Inagaki, and Sylvia Morelli for making graduate school a joy. Thanks to Diana Tamir, Jenn Pfeifer, Bill Welser, Bob Hornik, Joe Cappella, David Lydon-Staley, Andy Tan, Dolores Albarracin, Kevin Ochsner, Nick Valentino, Joe Kable, and Kathleen

Hall Jamieson for conversations that shaped my thinking as I wrote. Thanks to Sarah Banet-Weiser, John Jackson, and Michael Delli Carpini for support as deans of the Annenberg School for Communication. Thanks also to the incredible staff at Annenberg who make our research possible and supported this book in multiple ways.

Thanks to the funding agencies and program officers who have not only provided financial support to make our team's research possible but also asked questions that have pushed my thinking, introduced me to fantastic collaborators, and believed in the work, especially Becky Ferrer, Bill Klein, and Kelly Blake at the National Cancer Institute; Mary Kautz at the National Institute on Drug Abuse; Lynne Haverkos at the National Institute of Child Health and Human Development; Bill Casebeer, Adam Russell, Erica Briscoe, Victoria Romero, and Brian Ketler at the Defense Advanced Research Projects Agency; Edward Palazzolo and Fred Gregory at the Army Research Office; Jean Vettel and Javi Garcia at the Army Research Laboratory; Steve Cole and Jana Haritatos at HopeLab; and Bill Welser at Lotic.ai.

Several people let me interview them and provided thoughtful insights and compelling stories that shaped this book. I'm so grateful to: Jenny Slate, Ernie, Dan, Livia, and Nancy Grunfeld, Vic Strecher, Rachel Strecher, Tonya Mosley, Dani Bassett, Roland Seydel, Philip Hambach, Rick Kaplan, Raj Jayadev, and Lucas Johnson.

The current and former members of the Communication Neuroscience Lab have shaped the way I think about the problems in this book in so many ways and also did the research I describe from our lab: Jeesung Ahn, Mary Andrews, Elisa Baek, Joseph Bayer, Liz Beard, Chris Benitez, Shannon Burns, Taurean Butler, Melis Çakar, Josh Carp, José Carreras-Tartak, Chris Cascio, Hang-Yee Chan, Nicole Cooper, Jason Coronel, Dani Cosme, Bruce Doré, Boaz Hameiri, Susan Hao, Agnes Jasinska, Darin Johnson, Mia Jovanova, Yoona Kang, Minji Kim, Elissa Kranzler, Nina Lauharatanahirun, Lynda Lin, Jiaying Liu, Sicilia "Lolo" Lomax, Thandi Lyew, Kirsten Lydic, Rebecca Martin, Bradley Mattan, Steven Mesquiti, Samantha Moore-

Berg, Ben Muzekari, Laetitia Mwilambwe-Tshilobo, J. P. Obley, Matt O'Donnell, Prateekshit "Kanu" Pandey, Ally Paul, Jacob Pearl, Teresa Pegors, Rui Pei, Diego Reinero, Anthony Resnick, Keana Richards, Farah Sayed, Ralf Schmälzle, Christin Scholz, Kristin Shumaker, Sebastian Speer, Allie Sinclair, Frank Tinney, Steven Tompson, Omaya Torres, Nick Wasylyshyn, and Evan Wilkerson. Special thanks to Dani Cosme and Nicole Cooper for reading early drafts, to Steven Mesquiti and Ally Paul for helping me figure out how to have time to write a book, and to Steven for helping me collect data for the subtitle.

Graham Griffith helped me think about stories and how to tell them, introduced me to people and ideas that helped the science come alive, and continues to be a wonderful friend and source of inspiration. Thanks also to Terhi Nurminen, Xandro Xu, and the students in my doctoral seminar at Annenberg for helpful conversations about early chapter drafts.

When I first started this project, Angela Duckworth and Katy Milkman introduced me to book agents and have each provided ongoing advice and support throughout the process. The two of you are responsible for so many more behavioral scientists feeling that it is possible to communicate in this way. Thank you for that service and for being so generous with your time and talent. Thanks also to Jamil Zaki, Adam Grant, Ellen Langer, Carol Tell, and Michelle Segar for conversations about book writing.

Thanks to my family for making it possible and more fun to do this project. Thanks to my mom, Katherine, my first, most consistent and committed cheerleader, the person who most strongly emphasized the importance of communication throughout my life and provided input on the manuscript; my dad, Nils, who loved reading and showed me how patience makes it possible to build beautiful things; my grandma, Beverly, who read many drafts of the manuscript and has been an inspiration and support, always; additional thanks to my mom and grandma, as well as Grandma Nancy and Grandma Ell, for taking care of our kids during COVID and allowing me to write.

Thanks to my brother, Eric, and my sister, Lily, for being the best siblings in every way, including giving me feedback on this book, and to their partners, Laura and Baxter, as well as Linda Davis and Melissa Williams for thoughtful conversations and early feedback on chapters. Thanks to the rest of our Massachusetts family, including Dave, Michelle, and Audrey. I am filled with gratitude to the Powels, who welcomed us during the pandemic, helped take care of our kids, and kept us centered; Sam, Saima, and Alia brightened our days with love; Rebecca Saxe made me feel seen and understood; and D. J. Kovacs, Ben Shattuck, and Jenny Slate provided wise advice throughout the process. Thanks also to Eva, Mary Ann, Chris, and the McMahons and Burneses for grounding and thanks to the fairies for inspiration.

Alaina Mauro, Emma Fletcher, Anna Resek Chung, Ashley Turner Snyder, Niaz Karim, Natasha Jategaonkar, Alex Cohen, and Clara Weishahn provided phone support throughout the pandemic and the writing process. Olivia Roth read full versions of the manuscript and gave detailed comments and suggestions to make the book more accessible. Corinne Low and Sylvia Houghteling helped me stay calm and strong. I am deeply grateful for your friendship.

Thanks to Emmett and Theo, my sweet, funny, thoughtful kids, and Brett, the best partner anyone could ask for. Brett read many drafts, brainstormed with me, and made the book, and our life, better. Thanks to all three of you for all of the love and happiness.

# NOTES

## Introduction

xi **amount of reward your brain expects:** Bartra, Oscar, McGuire, Joseph T., and Kable, Joseph W. 2013. "The valuation system: A coordinate-based meta-analysis of BOLD fMRI experiments examining neural correlates of subjective value." *NeuroImage* 76: 412–27; Kable, Joseph W., and Glimcher, Paul W. 2009. "The neurobiology of decision: Consensus and controversy." *Neuron* 63(6): 733–45; Haber, Suzanne N., and Knutson, Brian. 2010. "The reward circuit: Linking primate anatomy and human imaging." *Neuropsychopharmacol.* 35(1): 4–26.

xii **A reward can be money, but it can also be friendship:** Izuma, Keise, Saito, Daisuke N., and Sadato, Norihiro. 2008. "Processing of social and monetary rewards in the human striatum." *Neuron* 58(2): 284–94.

xii **to finally run a marathon:** Morelli, Sylvia A., Knutson, Brian, and Zaki, Jamil. 2018. "Neural sensitivity to personal and vicarious reward differentially relates to prosociality and well-being." *Soc. Cogn. Affect. Neurosci.* 13(8): 831–39; Tricomi, Elizabeth, and Fiez, Julie A. 2008. "Feedback signals in the caudate reflect goal achievement on a declarative memory task." *NeuroImage* 41(3): 1154–67; Ming, Dan, Chen, Qunlin, Yang, Wenjing, Chen, Rui, Wei, Dongtao, et al. 2016. "Examining brain structures associated with the motive to achieve success and the motive to avoid failure: A voxel-based morphometry study." *Soc. Neurosci.* 11(1): 38–48.

xii **the brain can find reward:** Bartra, McGuire, and Kable, "The valuation system."

xii **wearing sunscreen and quitting smoking:** Falk, Emily B., Berkman, Elliot T., Mann, Traci, Harrison, Brittany, and Lieberman, Matthew D. 2010. "Predicting persuasion-induced behavior change from the brain." *J. Neurosci.* 30(25): 8421–24; Falk, Emily B., Berkman, Elliot T., and Lieberman, Matthew D. 2012. "From neural responses to population behavior: Neural focus group predicts population-level media effects." *Psychol. Sci.* 23(5): 439–45.

xii **exercise more and drive safely:** Kang, Yoona, Cooper, Nicole, Pandey, Prateekshit, Scholz, Christin, O'Donnell, Matthew Brook, et al. 2018. "Effects of self-transcendence on neural responses to persuasive messages and health behavior change." *Proc. Natl. Acad. Sci. USA* 115(40): 9974–79; Falk, Emily B., O'Donnell, Matthew Brook, Cascio, Christopher N., Tinney, Francis, Kang, Yoona, et al. 2015. "Self-affirmation alters the brain's response to health messages and subsequent behavior change." *Proc. Natl. Acad. Sci. USA* 112(7): 1977–82; Pei, Rui, Lauharatanahirun, Nina, Cascio, Christopher N., O'Donnell, Matthew B., Shope, Jean T., et al. 2020. "Neural processes during adolescent risky decision making are associated with conformity to peer influence." *Dev. Cogn. Neurosci.* 44: 100794.

xiii **deciding what to eat:** Hare, Todd A., Camerer, Colin F., and Rangel, Antonio. 2009. "Self-control in decision-making involves modulation of the vmPFC valuation system." *Science* 324(5927): 646–48; Plassmann, Hilke, O'Doherty, John, Shiv, Baba, and Rangel, Antonio. 2008. "Marketing actions can modulate neural representations of experienced pleasantness." *Proc. Natl. Acad. Sci. USA* 105(3): 1050–54; Plassmann, Hilke, O'Doherty, John, and Rangel, Antonio. 2007. "Orbitofrontal cortex encodes willingness to pay in everyday economic transactions." *J. Neurosci.* 27(37): 9984–88.

xiii **what to buy:** Knutson, Brian, Rick, Scott, Wimmer, G. Elliott, Prelec, Drazen, and Loewenstein, George. 2007. "Neural predictors of purchases." *Neuron* 53(1): 147–56.

xiii **how much to save:** Frydman, Cary, and Camerer, Colin F. 2016. "The psychology and neuroscience of financial decision making." *Trends Cogn. Sci.* 20(9): 661–75; Ersner-Hershfield, Hal, Wimmer, G. Elliott, and Knutson, Brian. 2009. "Saving for the future self: Neural measures of future self-continuity predict temporal discounting." *Soc. Cogn. Affect. Neurosci.* 4(1): 85–92; Kable, Joseph W., and Glimcher, Paul W. 2007. "The neural correlates of subjective value during intertemporal choice." *Nat. Neurosci.* 10(12): 1625–33.

xiii **and more:** Smidts, Ale, Hsu, Ming, Sanfey, Alan G., Boksem, Maarten A. S., Ebstein, Richard B., et al. 2014. "Advancing consumer neuroscience." *Mark. Lett.* 25(3): 257–67.

xiv **a key ingredient for happiness and well-being:** Cosme, Danielle, Mesquiti, Steven, Falk, Emily, and Burns, Shannon. Forthcoming. "Self-reflection interventions to promote well-being."

xiv **why they are doing what they are doing:** Scientists call it agentic alignment when we can align our choices with our goals and values. Morris, Adam, Carlson, Ryan W., Kober, Hedy, and Crockett, Molly. 2023. "Introspective access to value-based choice processes." *PsyArXiv.* Preprint.

xvi **ride their bikes more:** Blumberg, Alex, and Pierre-Louis, Kendra. "Make biking cool (again)!" In *How to Save a Planet*, podcast, 37: 39. August 18, 2022.

xvii **a crack for light to peek through:** Zurn, Perry, and Bassett, Dani S. 2022. *Curious Minds: The Power of Connection.* MIT Press.

xviii **a stronger sense of purpose:** Cosme, Mesquiti, Falk, and Burns, "Self-reflection interventions to promote well-being."

xix **the more they learned:** Nguyen, Mai, Chang, Ashley, Micciche, Emily, Meshulam, Meir, Nastase, Samuel A., and Hasson, Uri. 2021. "Teacher–student neural coupling during teaching and learning." *Soc. Cogn. Affect. Neurosci.* 17(4): 367–76.

xix **certain kinds of problem-solving tasks:** Reinero, Diego A., Dikker, Suzanne, and Van Bavel, Jay J. 2021. "Inter-brain synchrony in teams predicts collective performance." *Soc. Cogn. Affect. Neurosci.* 16(1–2): 43–57.

xix **People enjoy wide-ranging conversations:** Speer, S., Mwilambwe-Tshilobo, L., Tsoi, L., Burns, S., Falk, E., and Tamir, D. 2024. "What makes a good conversation? fMRI-hyperscanning shows friends explore and strangers converge." *Nat. Commun.* 15(1): 7781; Speer, S., Burns, S., Mwilambwe-Tshilobo, L., Tsoi, L., Falk, E., and Tamir, D. 2024. "Finding agreement: fMRI-hyperscanning reveals that dyads diverge in mental state space to align opinions." BioRxiv. Preprint.

xx **in a glass-bottom boat:** Tippett, Krista, and Johnson, Ayana Elizabeth. 2022. "Ayana Elizabeth Johnson—What if we get this right?" In *On Being*, podcast, 79: 34.

### A Note on the Research

xxi **where neurons are firing most heavily:** Huettel, Scott A., Song, Allen W., and McCarthy, Gregory. 2014. *Functional Magnetic Resonance Imaging*, 3rd ed. Sinauer Associates/Oxford University Press.

xxii **reconstruct the type of image:** Nishimoto, Shinji, Vu, An T., Naselaris, Thomas, Benjamini, Yuval, Yu, Bin, Gallant, Jack L. 2011. "Reconstructing visual experiences from brain activity evoked by natural movies." *Current biology* 21(19), 1641–46; Naselaris, Thomas, Prenger, Ryan J., Kay, Kendrick N., Oliver, Michael, and Gallant, Jack L. 2009. "Bayesian reconstruction of natural images from human brain activity." *Neuron* 63(6), 902–15.

xxii **white, Western, educated young adults:** Henrich, Joseph, Heine, Steven J., and Norenzayan, Ara. 2010. "The weirdest people in the world?" *Behav. Brain Sci.* 33(2–3): 61–83; discussion 83–135. But for a critique of the WEIRD acronym, see Syed, Moin, and Kathawalla, Ummul Kiram. 2021. "Cultural psychology, diversity, and representation in open science." In *Cultural Methods in Psychology: Describing and Transforming Cultures*, edited by Kate C. McLean, 427–54. Oxford University Press.

xxiii **across different cultures and contexts:** Falk, Emily B., Hyde, Luke W., Mitchell, Colter, Faul, Jessica, Gonzalez, Richard, et al. 2013. "What is a representative brain? Neuroscience meets population science." *Proc. Natl. Acad. Sci. USA* 110(44): 17615–22.

### Chapter 1: The Value Calculation

3 **"The People Hacker":** Unless otherwise noted, all quotes and story details about Jenny Radcliffe are from Rhysider, Jack, and Radcliffe, Jenny. 2021. "Ep 90: Jenny." In *Darknet Diaries*, podcast, audio, 69: 17, https://darknetdiaries.com/episode/90.

6 **scientists call the value system:** Kable, Joseph W., and Glimcher, Paul W. 2009. "The neurobiology of decision: Consensus and controversy." *Neuron* 63(6): 733–45; Bartra, Oscar, McGuire, Joseph T., and Kable, Joseph W. 2013. "The

valuation system: A coordinate-based meta-analysis of BOLD fMRI experiments examining neural correlates of subjective value." *NeuroImage* 76: 412–27.

8 **animal's well-being in the longer term:** Schultz, W. 1998. "Predictive reward signal of dopamine neurons." *J. Neurophysiol.* 80(1): 1–27.

8 **made them feel good:** Olds, James, and Milner, Peter. 1954. "Positive reinforcement produced by electrical stimulation of septal area and other regions of rat brain." *J. Comp. Physiol. Psychol.* 47(6): 419–27.

8 **food that they needed to stay alive:** Routtenberg, Aryeh, and Lindy, Janet. 1965. "Effects of the availability of rewarding septal and hypothalamic stimulation on bar pressing for food under conditions of deprivation." *J. Comp. Physiol. Psychol.* 60(2): 158–61.

8 **similar infrastructure in their brains:** Steiner, Jacob E., Glaser, Dieter, Hawilo, Maria E., and Berridge, Kent C. 2001. "Comparative expression of hedonic impact: Affective reactions to taste by human infants and other primates." *Neurosci. Biobehav. Rev.* 25(1): 53–74.

9 **researchers at Harvard Medical School:** Padoa-Schioppa, Camillo, and Assad, John A. 2006. "Neurons in the orbitofrontal cortex encode economic value." *Nature* 441(7090): 223–26.

11 **"reward" and "value" interchangeably:** Berkman, Elliot T., Livingston, Jordan L., and Kahn, Lauren E. 2017. "Finding the 'self' in self-regulation: The identity-value model." *Psychol. Inq.* 28(2–3): 77–98; Montague, P. Read, King-Casas, Brooks, and Cohen, Jonathan D. 2006. "Imaging valuation models in human choice." *Annu. Rev. Neurosci.* 29: 417–48.

12 **which choice Gizmo might make:** Padoa-Schioppa and Assad, "Neurons in the orbitofrontal cortex encode economic value."

12 **computing subjective values for each option:** Around the same time, other groups of scientists were finding similar patterns in regions of the monkey brain that had previously been thought to encode only basic reward and wanting that primates share with other mammals (e.g., hardwired responses to rewards that species need for survival, like food and sex). For example, they found that neurons in the striatum, a key part of the reward system in the center of the brain, also tracked the subjective value of choice options and the value of the option they ultimately chose. It was becoming increasingly clear that the system also kept track of a wider range of more complex features that give rise to more subjective notions of preference and value. Lau, Brian, and Glimcher, Paul W. 2008. "Value representations in the primate striatum during matching behavior." *Neuron* 58(3): 451–63.

12 **Hilke Plassmann and her colleagues at Caltech:** Plassmann, Hilke, O'Doherty, John, and Rangel, Antonio. 2007. "Orbitofrontal cortex encodes willingness to pay in everyday economic transactions." *J. Neurosci.* 27(37): 9984–88.

13 **track of the subjective value:** Plassmann, O'Doherty, and Rangel, "Orbitofrontal cortex encodes willingness to pay in everyday economic transactions."

13 **scientists at Caltech and Trinity College Dublin:** Chib, Vikram S., Rangel, Antonio, Shimojo, Shinsuke, and O'Doherty, John P. 2009. "Evidence for a

common representation of decision values for dissimilar goods in human ventromedial prefrontal cortex." *J. Neurosci.* 29(39): 12315–20.

14 **Around the same time, other groups:** Activity in the human ventral striatum and medial prefrontal cortex tracked people's willingness to pay different prices for a range of consumer goods. Knutson, Brian, Rick, Scott, Wimmer, G. Elliott, Prelec, Drazen, and Loewenstein, George. 2007. "Neural predictors of purchases." *Neuron* 53(1): 147–56.

14 **a 50 percent chance of winning $20:** Fox, Craig R., and Poldrack, Russell A. 2009. "Prospect theory and the brain." In *Neuroeconomics*, edited by Paul W. Glimcher, Colin F. Camerer, Ernst Fehr, and Russell A. Poldrack, 145–73. Academic Press.

14 **$10 now or $20 in six months?:** Kable, Joseph W., and Glimcher, Paul W. 2007. "The neural correlates of subjective value during intertemporal choice." *Nat. Neurosci.* 10(12): 1625–33.

14 **choices during the initial scan:** Levy, Ifat, Lazzaro, Stephanie C., Rutledge, Robb B., and Glimcher, Paul W. 2011. "Choice from non-choice: Predicting consumer preferences from blood oxygenation level-dependent signals obtained during passive viewing." *J. Neurosci.* 31(1): 118–25.

15 **things that aren't inherently comparable:** Levy, Dino J., and Glimcher, Paul W. 2012. "The root of all value: A neural common currency for choice." *Curr. Opin. Neurobiol.* 22(6): 1027–38.

16 **prediction and the actual outcome:** Ludwig, Vera U., Brown, Kirk Warren, and Brewer, Judson A. 2020. "Self-regulation without force: Can awareness leverage reward to drive behavior change?" *Perspect. Psychol. Sci.* 15(6): 1382–99.

16 **value calculation over time:** One important thing to notice, however, is that when we learn based on the outcome of a choice rather than focusing on the process that led us to choose, we sometimes come to the wrong conclusion about how good or bad that choice was. Former professional poker player and decision strategist Annie Duke's excellent and accessible summary in *How to Decide* offers many tools for improving our decision processes, to focus more on things we have control over, and reduce bias in our decision making. If you pause and think about what your best or worst choice this week was, you'll probably come up with choices that resulted in the best or worst outcomes. This *resulting* or *outcome bias* means that we sometimes learn the wrong lesson, since the outcome is partly shaped by our choice process and partly by luck; if you bike to the grocery store without a helmet and happen to make it home without a head injury, or if you ask your crush to the prom and they happen to have just said yes to someone else, that doesn't mean you shouldn't wear helmets or take chances on love. Sometimes good decisions turn out poorly and bad decisions turn out well, because of luck. Taking time to audit our choice process, independent of the outcome, can help overcome this bias. What information did you have before you made the choice? What other information could you have easily gotten? If you had it to do over again, what else would you want to take into account that you could have known before you made the

choice? By noticing that sometimes good decisions can turn out badly because of factors beyond our control, and bad decisions can turn out well by chance or luck, we can learn to learn better and improve our decision process, rather than fixating exclusively on the outcomes. Duke, Annie. 2020. *How to Decide: Simple Tools for Making Better Choices.* Penguin.

17 **value-based decision-making:** Kable and Glimcher, "The neurobiology of decision: Consensus and controversy."

17 **rather than improving our process:** Duke, *How to Decide.*

19 **UV rays can cause skin cancer:** Armstrong, Bruce K., and Kricker, Anne. 2001. "The epidemiology of UV induced skin cancer." *J. Photochem. Photobiol. B* 63(1–3): 8–18.

20 **exercise they got later:** Kang, Yoona, Cooper, Nicole, Pandey, Prateekshit, Scholz, Christin, O'Donnell, Matthew Brook, et al. 2018. "Effects of self-transcendence on neural responses to persuasive messages and health behavior change." *Proc. Natl. Acad. Sci. USA* 115(40): 9974–79; Falk, Emily B., O'Donnell, Matthew B., Cascio, Christopher N., Tinney, Francis, Kang, Yoona, et al. 2015. "Self-affirmation alters the brain's response to health messages and subsequent behavior change." *Proc. Natl. Acad. Sci. USA* 112(7): 1977–82.

20 **smoking over the following month:** Falk, Emily B., Berkman, Elliot T., Whalen, Danielle, and Lieberman, Matthew D. 2011. "Neural activity during health messaging predicts reductions in smoking above and beyond self-report." *Health Psychol.* 30(2): 177–85; Cooper, Nicole, Tompson, Steve, O'Donnell, Matthew Brook, and Falk, Emily B. 2015. "Brain activity in self- and value-related regions in response to online antismoking messages predicts behavior change." *J. Media Psychol.* 27: 93–109.

20 **brain responses and self-report:** Importantly, observing what was going on in the smokers' brains significantly improved our predictions of how much people would reduce their smoking in the following month. Knowing how much activation increased in the brain's value system when people watched anti-smoking ads made us better at guessing who would reduce their smoking, compared with when we only had access to self-reported survey answers about their confidence about their ability to change or their intention to change. It was better to have both types of measures than either one alone.

20 **captured by surveys alone:** Robins, Richard W., Fraley, R. Chris, and Krueger, Robert F. 2009. *Handbook of Research Methods in Personality Psychology.* Guilford Press; Nisbett, Richard E., and Wilson, Timothy D. 1977. "Telling more than we can know: Verbal reports on mental processes." *Psychol. Rev.* 84(3): 231–59.

20 **habitual routines:** If you are interested in exploring the kinds of behaviors that are typically guided by habits and the brain systems that support habitual behavior, there are many excellent books on these topics. I like Wendy Wood's *Good Habits, Bad Habits* and Russ Poldrack's *Hard to Break.* Wood, Wendy. 2019. *Good Habits, Bad Habits: The Science of Making Positive Changes That Stick.* Farrar, Straus and Giroux; Poldrack, Russell A. 2022. *Hard to Break: Why Our Brains Make Habits Stick.* Princeton University Press.

21 **why we do what we do:** Morris, Adam, Carlson, Ryan W., Kober, Hedy, and

Crockett, Molly. 2023. "Introspective access to value-based choice processes." PsyArXiv. Preprint.

22 **memory systems:** Wimmer, G. Elliott, and Shohamy, Daphna. 2012. "Preference by association: How memory mechanisms in the hippocampus bias decisions." *Science* 338(6104): 270–73.

22 **attention systems:** Smith, Stephanie M., and Krajbich, Ian. 2019. "Gaze amplifies value in decision making." *Psychol. Sci.* 30(1): 116–28.

22 **reasoning and regulating our emotions:** Hare, Todd A., Camerer, Colin F., and Rangel, Antonio. 2009. "Self-control in decision-making involves modulation of the vmPFC valuation system." *Science* 324(5927): 646–48.

23 **comes from the *social relevance system*:** I'm using the term "social relevance" to include a range of more specific functions that include the brain's ability to understand what people think and feel (which scientists call "mentalizing" or "theory of mind") and make use of that information to predict what people might think, feel, and do in the future. Social relevance also encompasses the brain's ability to assess what other people like and don't like and to make sense of how other people's thoughts and feelings (including their preferences) might affect us.

### *Chapter 2:* Who Am I?

26 **"How do I seem like":** Unless otherwise noted, all quotes come from Jenny Slate in personal correspondence with the author, August 2022.

26 **the problem was *her*:** Carey, Emma. 2022. "How Jenny Slate found her voice through Marcel the Shell." *Esquire,* June 28, 2022.

28 **how the brain tracks self-relevance:** Jenkins, Adrianna C., and Mitchell, Jason P. 2011. "Medial prefrontal cortex subserves diverse forms of self-reflection." *Soc. Neurosci.* 6(3): 211–18.

29 **these different parts of ourselves:** Elder, Jacob, Cheung, Bernice, Davis, Tyler, and Hughes, Brent. 2023. "Mapping the self: A network approach for understanding psychological and neural representations of self-concept structure." *J. Pers. Soc. Psychol.* 124(2): 237–63.

29 **imagine the future:** Spreng, R. Nathan, Mar, Raymond A., and Kim, Alice S. N. 2009. "The common neural basis of autobiographical memory, prospection, navigation, theory of mind, and the default mode: A quantitative meta-analysis." *J. Cogn. Neurosci.* 21(3): 489–510.

30 **significance of the memory:** D'Argembeau, Arnaud, Cassol, Helena, Phillips, Christophe, Balteau, Evelyne, Salmon, Eric, and Van der Linden, Martial. 2014. "Brains creating stories of selves: The neural basis of autobiographical reasoning." *Soc. Cogn. Affect. Neurosci.* 9(5): 646–52.

31 **different parts of the self-relevance system:** D'Argembeau, Cassol, Phillips, Balteau, Salmon, and Van der Linden, "Brains creating stories of selves."

31 **thinking about the future:** Spreng, Mar, and Kim, "The common neural basis of autobiographical memory, prospection, navigation, theory of mind, and the default mode: A quantitative meta-analysis."

31 **experience the present as the most vivid:** Lee, Sangil, Parthasarathi, Trishala, Cooper, Nicole, Zauberman, Gal, Lerman, Caryn, and Kable, Joseph W. 2022. "A neural signature of the vividness of prospective thought is modulated by temporal proximity during intertemporal decision making." *Proc. Natl. Acad. Sci. USA* 119(44): e2214072119.

32 **overlaps heavily with the value system:** Chavez, Robert S., Heatherton, Todd F., and Wagner, Dylan D. 2017. "Neural population decoding reveals the intrinsic positivity of the self." *Cereb. Cortex* 27(11): 5222–29.

33 **brain patterns that tracked value:** Mattan, Bradley, Cooper, Nicole, Scholz, Christin, Kang, Yoona, and Falk, Emily. 2021. "Neural signatures differentiating self-relevance and valence predict receptivity and adherence to health messages." Poster presented at the Annual Conference of the Social and Affective Neuroscience Society, Virtual, April 30, 2021.

33 **themselves versus others:** Chavez, Heatherton, and Wagner, "Neural population decoding reveals the intrinsic positivity of the self."

34 **a performance of who we are:** Gibbs, Jennifer L., Ellison, Nicole B., and Heino, Rebecca D. 2006. "Self-presentation in online personals: The role of anticipated future interaction, self-disclosure, and perceived success in internet dating." *Communic. Res.* 33(2): 152–77; Ellison, Nicole, Heino, Rebecca, and Gibbs, Jennifer. 2006. "Managing impressions online: Self-presentation processes in the online dating environment." *J. Comput. Mediat. Commun.* 11(2): 415–41.

34 **ourselves as "above average":** Dunning, David, Heath, Chip, and Suls, Jerry M. 2005. "Picture imperfect." *Sci. Am. Mind* 16(4): 20–27.

34 **personality traits described them:** Alicke, Mark D. 1985. "Global self-evaluation as determined by the desirability and controllability of trait adjectives." *J. Pers. Soc. Psychol.* 49(6): 1621–30.

35 **compared with generic ones:** Noar, Seth M., Benac, Christina N., and Harris, Melissa S. 2007. "Does tailoring matter? Meta-analytic review of tailored print health behavior change interventions." *Psychol. Bull.* 133(4): 673–93; Falk, Emily, Scholz, Christin, Cooper, Nicole, and Gomes, Victor. In press. "Social influence and behavior change." In *Handbook of Social Psychology*, 6th ed., edited by Daniel Gilbert, Susan Fiske, Eli Finkel, and Wendy Mendes. Wiley.

36 **18 percent in the control group:** Strecher, Victor J., Shiffman, Saul, and West, Robert. 2005. "Randomized controlled trial of a web-based computer-tailored smoking cessation program as a supplement to nicotine patch therapy." *Addiction* 100(5): 682–88.

36 **compared with generic messaging:** Chua, Hannah Faye, Ho, S. Shaun, Jasinska, Agnes J., Polk, Thad A., Welsh, Robert C., et al. 2011. "Self-related neural response to tailored smoking-cessation messages predicts quitting." *Nat. Neurosci.* 14(4): 426–27.

36 **medial prefrontal cortex and precuneus:** Chua, Ho, Jasinska, Polk, Welsh, et al., "Self-related neural response to tailored smoking-cessation messages predicts quitting."

37 **a backpack might make them feel:** Aquino, Antonio, Alparone, Francesca

Romana, Pagliaro, Stefano, Haddock, Geoffrey, Maio, Gregory R., et al. 2020. "Sense or sensibility? The neuro-functional basis of the structural matching effect in persuasion." *Cogn. Affect. Behav. Neurosci.* 20(3): 536–50.

37 **The reverse is true:** Aquino, Alparone, Pagliaro, Haddock, Maio, et al., "Sense or sensibility?"

37 **changes to our day-to-day behavior:** Berkman, Elliot T., Livingston, Jordan L., and Kahn, Lauren E. 2017. "Finding the 'self' in self-regulation: The identity-value model." *Psychol. Inq.* 28(2–3): 77–98.

38 **incompatible with who we are:** Langer, Ellen J. 2014. *Mindfulness.* Hachette Books. 25th anniversary ed.; Maymin, Philip Z., and Langer, Ellen J. 2021. "Cognitive biases and mindfulness." *Humanit. Soc. Sci.* 8(1): 1–11.

39 **identity-value model of self-control:** Berkman, Livingston, and Kahn, "Finding the 'self' in self-regulation: The identity-value model."

40 **also correct and good:** Couldry, Nick, and Turow, Joseph. 2014. "Advertising, big data and the clearance of the public realm: Marketers' new approaches to the content subsidy." *Int. J. Commun. Syst.* 8: 1710–26; Albarracin, Dolores, Albarracin, Julia, Chan, Man Pui Sally, and Jamieson, Kathleen Hall. 2021. *Creating Conspiracy Beliefs: How Our Thoughts Are Shaped.* Cambridge University Press; Benjamin, Ruha. 2023. "Race after technology." In *Social Theory Re-Wired*, 3rd ed., edited by Wesley Longhofer and Daniel Winchester, 405–15. Routledge; Lazer, David M. J., Baum, Matthew A., Benkler, Yochai, Berinsky, Adam J., Greenhill, Kelly M., et al. 2018. "The science of fake news." *Science* 359(6380): 1094–96.

40 **to boost our self-esteem:** Elder, Cheung, Davis, and Hughes, "Mapping the self"; Taylor, Shelley E. 1989. *Positive Illusions: Creative Self-Deception and the Healthy Mind.* Basic Books.

42 **so tightly to them:** Langer, *Mindfulness*; Frewen, Paul A., Evans, Elspeth M., Maraj, Nicholas, Dozois, David J. A., and Partridge, Kate. 2008. "Letting go: Mindfulness and negative automatic thinking." *Cognit. Ther. Res.* 32(6): 758–74.

43 **no longer being separate and unique:** Sacchet, Michael D., and Brewer, Judson A. 2024. "Beyond mindfulness: Advanced meditation alters consciousness and our basic sense of self." *Sci. Amer.* 331(1, June/July): 70.

43 **part of the brain than nonmeditators:** Jang, Joon Hwan, Jung, Wi Hoon, Kang, Do-Hyung, Byun, Min Soo, Kwon, Soo Jin, et al. 2011. "Increased default mode network connectivity associated with meditation." *Neurosci. Lett.* 487(3): 358–62.

43 **communicates with other brain regions:** Farb, Norman A. S., Segal, Zindel V., Mayberg, Helen, Bean, Jim, McKeon, Deborah, et al. 2007. "Attending to the present: Mindfulness meditation reveals distinct neural modes of self-reference." *Soc. Cogn. Affect. Neurosci.* 2(4): 313–22.

43 **restricted sense of self:** Gattuso, James J., Perkins, Daniel, Ruffell, Simon, Lawrence, Andrew J., Hoyer, Daniel, et al. 2023. "Default mode network modulation by psychedelics: A systematic review." *Int. J. Neuropsychopharmacol.* 26(3): 155–88.

43 **connectedness to other people:** Forstmann, Matthias, Yudkin, Daniel A.,

Prosser, Annayah M. B., Heller, S. Megan, and Crockett, Molly J. 2020. "Transformative experience and social connectedness mediate the mood-enhancing effects of psychedelic use in naturalistic settings." *Proc. Natl. Acad. Sci. USA* 117(5): 2338–46.

45 **what other people see in us:** Pfeifer, Jennifer H., Masten, Carrie L., Borofsky, Larissa A., Dapretto, Mirella, Fuligni, Andrew J., and Lieberman, Matthew D. 2009. "Neural correlates of direct and reflected self-appraisals in adolescents and adults: When social perspective-taking informs self-perception." *Child Dev.* 80(4): 1016–38; Pfeifer, Jennifer H., Mahy, Caitlin E. V., Merchant, Junaid S., Chen, Chunhui, Masten, Carrie L., et al. 2017. "Neural systems for reflected and direct self-appraisals in Chinese young adults: Exploring the role of the temporal-parietal junction." *Cultur. Divers. Ethnic Minor. Psychol.* 23(1): 45–58; Van der Cruijsen, Renske, Peters, Sabine, Zoetendaal, Kelly P. M., Pfeifer, Jennifer H., and Crone, Eveline A. 2019. "Direct and reflected self-concept show increasing similarity across adolescence: A functional neuroimaging study." *Neuropsychologia* 129: 407–17; Pfeifer, Jennifer H., and Peake, Shannon J. 2012. "Self-development: Integrating cognitive, socioemotional, and neuroimaging perspectives." *Dev. Cogn. Neurosci.* 2(1): 55–69.

45 **to figure out who we are:** Wallace, Harry M., and Tice, Dianne M. 2012. "Reflected appraisal through a 21st-century looking glass." In M. R. Leary and J. P. Tangney, eds., *Handbook of Self and Identity*, 2nd ed., 124–140. Guilford Press.

### Chapter 3: Who Are We?

47 *Glamour* **magazine has:** Turner, Aidan. 2016. "GLAMOUR 100 Sexiest Men 2016." *Glamour Magazine UK*, February 1, 2016.

48 **"It just makes me giggle":** Radio Absolute. 2013. Benedict Cumberbatch on being "The Sexiest Man Alive." https://www.youtube.com/watch?v=IZUiA_oELOs.

48 **repeatedly mispronounced "penguins":** BBC America. 2014. Benedict Cumberbatch Can't Say "Penguins." *The Graham Norton Show on BBC America*.

49 **over two hundred other women's faces:** Klucharev, Vasily, Hytönen, Kaisa, Rijpkema, Mark, Smidts, Ale, and Fernández, Guillén. 2009. "Reinforcement learning signal predicts social conformity." *Neuron* 61(1): 140–51.

49 **See the endnote for more on why:** Bok, Sissela. 2011. "Deceptive social science research." In *Lying: Moral Choice in Public and Private Life*, 182–202. Vintage.

Although researchers have different views on the costs and usefulness of deception, all research studies with human participants that are done at universities are reviewed by an ethics board to weigh the costs and benefits (that is, is the deception justified and risks minimized). Importantly, in these studies, the research team tells the participants in advance that they will not receive complete information about the study, and people decide whether they want to consent to being in a study where they can't know everything in advance. This consent process includes an explanation that the study may involve some deception without revealing its nature. After the study is complete, the participants are typically given the correct information (they learn the truth). This involves explaining the true purpose of the study, the nature of the deception, and what purpose it served.

50 **exposed to the peer feedback:** Zaki, Jamil, Schirmer, Jessica, and Mitchell, Jason P. 2011. "Social influence modulates the neural computation of value." *Psychol. Sci.* 22(7): 894–900.

51 **From what foods we want to eat:** Nook, Erik C., and Zaki, Jamil. 2015. "Social norms shift behavioral and neural responses to foods." *J. Cogn. Neurosci.* 27(7): 1412–26.

51 **products we would recommend:** Cascio, Christopher N., O'Donnell, Matthew Brook, Bayer, Joseph, Tinney, Francis J., Jr., and Falk, Emily B. 2015. "Neural correlates of susceptibility to group opinions in online word-of-mouth recommendations." *J. Mark. Res.* 52(4): 559–75.

51 **display on our walls:** Welborn, B. Locke, Lieberman, Matthew D., Goldenberg, Diane, Fuligni, Andrew J., Galván, Adriana, and Telzer, Eva H. 2016. "Neural mechanisms of social influence in adolescence." *Soc. Cogn. Affect. Neurosci.* 11(1): 100–109.

51 **beliefs, and who we are:** Albarracín, Dolores. 2021. *Action and Inaction in a Social World: Predicting and Changing Attitudes and Behavior.* Cambridge University Press.

51 **whether to vote:** Bond, Robert M., Fariss, Christopher J., Jones, Jason J., Kramer, Adam D. I., Marlow, Cameron, et al. 2012. "A 61-million-person experiment in social influence and political mobilization." *Nature* 489(7415): 295–98.

51 **pay taxes:** Hallsworth, Michael, List, John A., Metcalfe, Robert D., and Vlaev, Ivo. 2017. "The behavioralist as tax collector: Using natural field experiments to enhance tax compliance." *J. Public Econ.* 148: 14–31.

51 **exercise:** Zhang, Jingwen, Brackbill, Devon, Yang, Sijia, and Centola, Damon. 2015. "Efficacy and causal mechanism of an online social media intervention to increase physical activity: Results of a randomized controlled trial." *Prev. Med. Rep.* 2: 651–57.

51 **decision-making and that of others:** Cialdini, Robert B., and Goldstein, Noah J. 2004. "Social influence: Compliance and conformity." *Annu. Rev. Psychol.* 55: 591–621.

52 **others might think and feel:** Saxe, Rebecca, and Kanwisher, Nancy. 2003. "People thinking about thinking people: The role of the temporo-parietal junction in 'theory of mind.'" *NeuroImage* 19(4): 1835–42.

52 **how it might affect us:** Thornton, Mark A., Weaverdyck, Miriam E., and Tamir, Diana I. 2019. "The social brain automatically predicts others' future mental states." *J. Neurosci.* 39(1): 140–48; Thornton, Mark A., and Tamir, Diana I. 2021. "People accurately predict the transition probabilities between actions." *Sci. Adv.* 7(9): eabd4995; Thornton, Mark A., and Tamir, Diana I. 2017. "Mental models accurately predict emotion transitions." *Proc. Natl. Acad. Sci. USA* 114(23): 5982–87; Tamir, Diana I., and Thornton, Mark A. 2018. "Modeling the predictive social mind." *Trends Cogn. Sci.* 22(3): 201–12.

52 **other types of inferences:** Saxe and Kanwisher, "People thinking about thinking people."

54 **people on the autism spectrum:** Lombardo, Michael V., Chakrabarti, Bhismadev, Bullmore, Edward T., MRC AIMS Consortium, and Baron-Cohen, Simon. 2011. "Specialization of right temporo-parietal junction for mentalizing and its relation to social impairments in autism." *NeuroImage* 56(3): 1832–38.

54 **associated with social difficulties:** Chien, Hsiang-Yun, Lin, Hsiang-Yuan, Lai, Meng-Chuan, Gau, Susan Shur-Fen, and Tseng, Wen-Yih Isaac. 2015. "Hyperconnectivity of the right posterior temporo-parietal junction predicts social difficulties in boys with autism spectrum disorder." *Autism Res.* 8(4): 427–41.

55 **when something unexpected happens:** Koster-Hale, Jorie, and Saxe, Rebecca. 2013. "Theory of mind: A neural prediction problem." *Neuron* 79(5): 836–48.

55 **behaviors might follow another:** Thornton and Tamir, "People accurately predict the transition probabilities between actions."

55 **they heard the story the first time:** Baldassano, Christopher, Chen, Janice, Zadbood, Asieh, Pillow, Jonathan W., Hasson, Uri, and Norman, Kenneth A. 2017. "Discovering event structure in continuous narrative perception and memory." *Neuron* 95(3): 709–21.

56 **short film, *Partly Cloudy*, twice:** Richardson, Hilary, and Saxe, Rebecca. 2020. "Development of predictive responses in theory of mind brain regions." *Dev. Sci.* 23(1): e12863.

57 **I Want You to Want Me:** (Cheap Trick song title) Nielsen, Rick. 1977. *In Color.* Epic.

57 **humanity's deep past:** Dunbar, R. I. M., and Shultz, Susanne. 2007. "Evolution in the social brain." *Science* 317(5843): 1344–47.

58 **industries dominated by men:** Milestone, Katie. 2015. "Gender and the cultural industries." In *The Routledge Companion to the Cultural Industries*, edited by Kate Oakley and Justin O'Connor, 501–11. Routledge; Döring, Nicola, and Mohseni, M. Rohangis. 2019. "Male dominance and sexism on YouTube: Results of three content analyses." *Fem. Media Stud.* 19(4): 512–24; Krijnen, Tonny, and Van Bauwel, Sofie. 2021. *Gender and Media: Representing, Producing, Consuming.* Routledge; Lauzen, Martha M., and Dozier, David M. 2004. "Evening the score in prime time: The relationship between behind-the-scenes women and on-screen portrayals in the 2002–2003 season." *J. Broadcast. Electron. Media* 48(3): 484–500; Chemaly, Soraya. 2019. "Demographics, design, and free speech: How demographics have produced social media optimized for abuse and the silencing of marginalized voices." *Free Speech in the Digital Age*, edited by Susan J. Brison and Katharine Gelber, 150–69.

58 **to repair our social ties:** Eisenberger, Naomi I. 2015. "Social pain and the brain: Controversies, questions, and where to go from here." *Annu. Rev. Psychol.* 66: 601–29.

59 **like chocolate and money:** Izuma, Keise, Saito, Daisuke N., and Sadato, Norihiro. 2008. "Processing of social and monetary rewards in the human striatum." *Neuron* 58(2): 284–94; Lieberman, Matthew D., and Eisenberger, Naomi I. 2009. "Neuroscience. Pains and pleasures of social life." *Science* 323(5916): 890–91.

59 **activating feelings of pleasure:** Meier, Isabell M., Eikemo, Marie, and Leknes, Siri. 2021. "The role of mu-opioids for reward and threat processing in humans: Bridging the gap from preclinical to clinical opioid drug studies." *Curr. Addict. Rep.* 8(2): 306–18.

59 **you connect with loved ones:** Meier, Eikemo, and Leknes, "The role of mu-opioids for reward and threat processing in humans."

59 **and make us feel good:** Inagaki, Tristen K., Hazlett, Laura I., and Andreescu, Carmen. 2020. "Opioids and social bonding: Effect of naltrexone on feelings of social connection and ventral striatum activity to close others." *J. Exp. Psychol. Gen.* 149(4): 732–45.

60 **influencing their behavior:** Dallas, Rebecca, Field, Matt, Jones, Andrew, Christiansen, Paul, Rose, Abi, and Robinson, Eric. 2014. "Influenced but unaware: Social influence on alcohol drinking among social acquaintances." *Alcohol. Clin. Exp. Res.* 38(5): 1448–53; Nolan, Jessica M., Schultz, P. Wesley, Cialdini, Robert B., Goldstein, Noah J., and Griskevicius, Vladas. 2008. "Normative social influence is underdetected." *Pers. Soc. Psychol. Bull.* 34(7): 913–23.

60 **benefiting society and the environment:** Nolan, Schultz, Cialdini, Goldstein, and Griskevicius, "Normative social influence is underdetected."

61 **received other kinds of messages:** Nolan, Schultz, Cialdini, Goldstein, and Griskevicius, "Normative social influence is underdetected."

61 **the least important reason:** Nolan, Schultz, Cialdini, Goldstein, and Griskevicius, "Normative social influence is underdetected."

61 **from towel reuse in hotels:** Goldstein, Noah J., Cialdini, Robert B., and Griskevicius, Vladas. 2008. "A room with a viewpoint: Using social norms to motivate environmental conservation in hotels." *J. Consum. Res.* 35(3): 472–82.

61 **to voting:** Bond, Fariss, Jones, Kramer, Marlow, et al., "A 61-million-person experiment in social influence and political mobilization."

61 **to exercise:** Zhang, Brackbill, Yang, and Centola, "Efficacy and causal mechanism of an online social media intervention to increase physical activity."

61 **affecting their opinions and actions:** Dallas, Field, Jones, Christiansen, Rose, and Robinson, "Influenced but unaware."

61 **not more effective than standard messaging:** Bohner, Gerd, and Schlüter, Lena E. 2014. "A room with a viewpoint revisited: Descriptive norms and hotel guests' towel reuse behavior." *PLoS One* 9(8): e104086.

62 **coached about getting more exercise:** Pandey, Prateekshit, Kang, Yoona, Cooper, Nicole, O'Donnell, Matthew B., and Falk, Emily B. 2021. "Social networks and neural receptivity to persuasive health messages." *Health Psychol.* 40(4): 285–94; Zhang, Jingwen, Brackbill, Devon, Yang, Sijia, Becker, Joshua, Herbert, Natalie, and Centola, Damon. 2016. "Support or competition? How online social networks increase physical activity: A randomized controlled trial." *Prev. Med. Rep.* 4: 453–58.

63 **social relevance, social support:** Milsom, Vanessa A., Perri, Michael G., and Rejeski, W. Jack. 2007. "Guided group support and the long-term management of obesity." In *Self-Help Approaches for Obesity and Eating Disorders: Research*

*and Practice*, edited by Janet D. Latner and G. Terence Wilson, 205–22. Guilford Press; Renjilian, David A., Perri, Michael G., Nezu, Arthur M., McKelvey, Wendy F., Shermer, Rebecca L., and Anton, Stephen D. 2001. "Individual versus group therapy for obesity: Effects of matching participants to their treatment preferences." *J. Consult. Clin. Psychol.* 69(4): 717–21; Wing, Rena R., and Jeffery, Robert W. 1999. "Benefits of recruiting participants with friends and increasing social support for weight loss and maintenance." *J. Consult. Clin. Psychol.* 67(1): 132–38; Ahn, Jeesung, Falk, Emily B., and Kang, Yoona. 2024. "Relationships between physical activity and loneliness: A systematic review of intervention studies." *Curr. Res. Behav. Sci.* 6: 100141.

63 **and commitment:** Lokhorst, Anne Marike, Werner, Carol, Staats, Henk, van Dijk, Eric, and Gale, Jeff L. 2013. "Commitment and behavior change: A meta-analysis and critical review of commitment-making strategies in environmental research." *Environ. Behav.* 45(1): 3–34; Baca-Motes, Katie, Brown, Amber, Gneezy, Ayelet, Keenan, Elizabeth A., and Nelson, Leif D. 2012. "Commitment and behavior change: Evidence from the field." *J. Consum. Res.* 39(5): 1070–84.

64 **The Emperor's New Clothes:** Folktale by Hans Christian Andersen.

64 **trails and taking them home:** Cialdini, Robert B., Demaine, Linda J., Sagarin, Brad J., Barrett, Daniel W., Rhoads, Kelton, and Winter, Patricia L. 2006. "Managing social norms for persuasive impact." *Social Influence* 1(1): 3–15.

64 **actually be part of the problem:** Cialdini, Demaine, Sagarin, Barrett, Rhoads, and Winter, "Managing social norms for persuasive impact."

64 **less likely to go for my run:** Aral, Sinan, and Nicolaides, Christos. 2017. "Exercise contagion in a global social network." *Nat. Commun.* 8: 14753.

65 **might become self-perpetuating:** Richards, Keana S. 2022. "Women Prepare More Than Men in Competitive and Non-competitive Environments, Which Aligns with Gender Stereotypes." PhD thesis. University of Pennsylvania.

67 **went along with the group:** Asch, Solomon E. 1956. "Studies of independence and conformity: I. A minority of one against a unanimous majority." *Psychol. Monogr.* 70(9): 1–70.

67 **faster online than true news:** Vosoughi, Soroush, Roy, Deb, and Aral, Sinan. 2018. "The spread of true and false news online." *Science* 359(6380): 1146–51; Watts, Duncan J., Rothschild, David M., and Mobius, Markus. 2021. "Measuring the news and its impact on democracy." *Proc. Natl. Acad. Sci. USA* 118(15): e1912443118.

68 **"Presenting partial or biased data":** Watts, Rothschild, and Mobius, "Measuring the news and its impact on democracy."

68 **weaponization of social norms:** Shao, Chengcheng, Ciampaglia, Giovanni Luca, Varol, Onur, Yang, Kai Cheng, Flammini, Alessandro, and Menczer, Filippo. 2018. "The spread of low-credibility content by social bots." *Nat. Commun.* 9(1): 4787.

68 **coordinated attacks from trolls:** Neilson, Tai, and Ortiga, Kara. 2023. "Mobs, crowds, and trolls: Theorizing the harassment of journalists in the Philippines." *Digit. Journal* 11(10): 1924–39.

68 **take these issues on alone:** Watts, Rothschild, and Mobius, "Measuring the news and its impact on democracy."

## Chapter 4: To Change What You Think, Change What You Think About

73 **Change What You Think About:** This idea stems from McCombs, Maxwell, and Shaw, Donald. 1972. "The agenda-setting function of mass media." *Public Opinion Quarterly* 36(2): 176–87. Bernard Cohen famously wrote that the media "may not be successful much of the time in telling people what to think, but it is stunningly successful in telling its readers what to think *about*." Cohen, Bernard. 1963. *The Press and Foreign Policy.* Princeton University Press.

73 **made-for-TV movie:** Wyche, Steve. 2003. "Grunfeld's Triumphant Journey: Romanian Born, N.Y. Bred, Veteran GM Looks to Remake Wizards." *Washington Post*, December 25, 2003.

73 **America's premier basketball league:** Allen, Scott. 2022. "Ernie Grunfeld's son details his family's journey from the Holocaust to the NBA in new book." *Washington Post*, January 7, 2022.

73 **get the family organized:** Grunfeld, Dan. 2021. *By the Grace of the Game: The Holocaust, a Basketball Legacy, and an Unprecedented American Dream.* Triumph Books.

74 **"Mrs. Grunfeld," the coach said:** Grunfeld, *By the Grace of the Game.*

75 **second all-time:** Ernie Grunfeld—Men's basketball. University of Tennessee Athletics. https://utsports.com/sports/mens-basketball/roster/ernie-grunfeld/8524.

77 **he had *fun*:** Grunfeld, *By the Grace of the Game.*

78 **traveling in time, space, or identity:** Tamir, Diana I., and Mitchell, Jason P. 2011. "The default network distinguishes construals of proximal versus distal events." *J. Cogn. Neurosci.* 23(10): 2945–55; see also Parkinson, Carolyn, Liu, Shari, and Wheatley, Thalia. 2014. "A common cortical metric for spatial, temporal, and social distance." *J. Neurosci.* 34(5): 1979–87.

78 **Right now, *I'm* having fun:** As comedian Jerry Seinfeld has observed: "I never get enough sleep. I stay up late at night 'cause I'm 'night guy.' Night guy wants to stay up late. 'What about getting up after five hours of sleep? Oh, that's morning guy's problem.'" Jerry Seinfeld. "The Glasses." Netflix video. 22:56. September 30, 1993. https://www.netflix.com/title/70153373.

78 **at different times in the future:** Kable, Joseph W., and Glimcher, Paul W. 2007. "The neural correlates of subjective value during intertemporal choice." *Nat. Neurosci.* 10(12): 1625–33.

79 **their future selves in their brains:** Ersner-Hershfield, Hal, Wimmer, G. Elliott, and Knutson, Brian. 2009. "Saving for the future self: Neural measures of future self-continuity predict temporal discounting." *Soc. Cogn. Affect. Neurosci.* 4(1): 85–92.

79 **abstract benefits and consequences:** Woolley, Kaitlin, and Fishbach, Ayelet. 2016. "For the fun of it: Harnessing immediate rewards to increase persistence in long-term goals." *J. Consum. Res.* 42(6): 952–66.

80 **choices like vegetables and salad:** Turnwald, Bradley P., and Crum, Alia J. 2019. "Smart food policy for healthy food labeling: Leading with taste, not healthiness, to shift consumption and enjoyment of healthy foods." *Prev. Med.* 119: 7–13.

81 **studying for her exams:** Milkman, Katy L. 2021. *How to Change: The Science of Getting from Where You Are to Where You Want to Be.* Vermillion.

81 **get more exercise as well:** Milkman, Katy L., Minson, Julia A., and Volpp, Kevin G. 2014. "Holding the Hunger Games hostage at the gym: An evaluation of temptation bundling." *Manage. Sci.* 60(2): 283–99.

82 **money to their retirement fund:** Hershfield, Hal E., Goldstein, Daniel G., Sharpe, William F., Fox, Jesse, Yeykelis, Leo, et al. 2011. "Increasing saving behavior through age-progressed renderings of the future self." *J. Mark. Res.* 48(SPL): S23–37.

83 **saving more for the future:** Bank of America Corporation. 2012. "Merrill Edge® Launches Face Retirement." Business Wire.

83 **patiently wait for them:** Ersner-Hershfield, Wimmer, and Knutson, "Saving for the future self."

83 **cravings for cigarettes and junk food:** Kober, Hedy, Mende-Siedlecki, Peter, Kross, Ethan F., Weber, Jochen, Mischel, Walter, et al. 2010. "Prefrontal-striatal pathway underlies cognitive regulation of craving." *Proc. Natl. Acad. Sci. USA* 107(33): 14811–16; Roos, Corey R., Harp, Nicholas R., Vafaie, Nilofar, Gueorguieva, Ralitza, Frankforter, Tami, et al. 2023. "Randomized trial of mindfulness- and reappraisal-based regulation of craving training among daily cigarette smokers." *Psychol. Addict. Behav.* 37(7): 829–40; Boswell, Rebecca G., Sun, Wendy, Suzuki, Shosuke, and Kober, Hedy. 2018. "Training in cognitive strategies reduces eating and improves food choice." *Proc. Natl. Acad. Sci. USA* 115(48): E11238–47.

83 **choices they ultimately made:** Hare, Todd A., Malmaud, Jonathan, and Rangel, Antonio. 2011. "Focusing attention on the health aspects of foods changes value signals in vmPFC and improves dietary choice." *J. Neurosci.* 31(30): 11077–87. Research led by Cendri Hutcherson and Anita Tusche at Caltech and the University of Toronto also shows that focusing on different choice attributes engages a common mechanism across different types of choices, including food choices and social decisions. Cendri and Anita found that the brain's value system tracks the overall value of different foods, but adds nuance to our understanding of how paying attention to different dimensions of a choice can change what we decide. Their research highlights how regions outside the core value system encode information about different goals (e.g., "eat healthier" when choosing foods to eat, or "be kinder" when making decisions about whether to share financial profits with others), which shape our choices. Tusche, Anita, and Hutcherson, Cendri A. 2018. "Cognitive regulation alters social and dietary choice by changing attribute representations in domain-general and domain-specific brain circuits." *eLife* 7: e31185.

84 **less healthy choice:** More general instructions to create distance between ourselves and the things we crave had similar effects, as have more general instructions to focus on things people liked about particular foods (e.g., how it tastes, how good it is for my body) or things they didn't like about different foods (e.g., how it tastes, how bad it is for my body). Boswell, Sun, Suzuki, and Kober, "Training in cognitive strategies reduces eating and improves food choice"; Hutcherson, Cendri A., Plassmann, Hilke, Gross, James J., and Ran-

gel, Antonio. 2012. "Cognitive regulation during decision making shifts behavioral control between ventromedial and dorsolateral prefrontal value systems." *J. Neurosci.* 32(39): 13543–54.

84 **impact cravings and behavior:** Cosme, Danielle, Zeithamova, Dagmar, Stice, Eric, and Berkman, Elliot T. 2020. "Multivariate neural signatures for health neuroscience: Assessing spontaneous regulation during food choice." *Soc. Cogn. Affect. Neurosci.* 15(10): 1120–34.

85 **misled during the decision process:** Chang, Linda W., and Cikara, Mina. 2018. "Social decoys: Leveraging choice architecture to alter social preferences." *J. Pers. Soc. Psychol.* 115(2): 206–23.

86 **bring diverse skills to the table:** Ruef, Martin, Aldrich, Howard E., and Carter, Nancy M. 2003. "The structure of founding teams: Homophily, strong ties, and isolation among U.S. entrepreneurs." *Am. Sociol. Rev.* 68(2): 195–222.

86 **idea novelty:** Stahl, Günter K., Maznevski, Martha L., Voigt, Andreas, and Jonsen, Karsten. 2010. "Unraveling the effects of cultural diversity in teams: A meta-analysis of research on multicultural work groups." *J. Int. Bus. Stud.* 41(4): 690–709; Antonio, Anthony Lising, Chang, Mitchell J., Hakuta, Kenji, Kenny, David A., Levin, Shana, and Milem, Jeffrey F. 2004. "Effects of racial diversity on complex thinking in college students." *Psychol. Sci.* 15(8): 507–10; Sommers, Samuel R. 2006. "On racial diversity and group decision making: Identifying multiple effects of racial composition on jury deliberations." *J. Pers. Soc. Psychol.* 90(4): 597–612; Chilton, Adam, Driver, Justin, Masur, Jonathan S., and Rozema, Kyle. 2022. "Assessing affirmative action's diversity rationale." *Columbia Law Rev.* 122(2): 331–406; Hofstra, Bas, Kulkarni, Vivek V., Galvez, Sebastian Munoz Najar, He, Bryan, Jurafsky, Dan, and McFarland, Daniel A. 2020. "The diversity–innovation paradox in science." *Proc. Natl. Acad. Sci. USA* 117(17): 9284–91.

86 **inequality across groups:** Craig, Maureen A., and Richeson, Jennifer A. 2014. "More diverse yet less tolerant? How the increasingly diverse racial landscape affects White Americans' racial attitudes." *Pers. Soc. Psychol. Bull.* 40(6): 750–61; Richeson, Jennifer A., and Sommers, Samuel R. 2016. "Toward a social psychology of race and race relations for the twenty-first century." *Annu. Rev. Psychol.* 67: 439–63; Richeson, Jennifer A., and Shelton, J. Nicole. 2007. "Negotiating interracial interactions: Costs, consequences, and possibilities." *Curr. Dir. Psychol. Sci.* 16(6): 316–20; Shelton, J. Nicole, Richeson, Jennifer A., and Salvatore, Jessica. 2005. "Expecting to be the target of prejudice: Implications for interethnic interactions." *Pers. Soc. Psychol. Bull.* 31(9): 1189–202.

87 **people from marginalized groups:** For instance, male students and white students were more likely to get to participate in internship programs, whereas first-generation college students and people receiving financial aid through federal Pell grants were less likely to get to participate. In one survey of college students, the majority of those who hadn't done internships wished they could have, citing the need to work at their main paid jobs as the most frequent barrier; indeed, those who worked fewer hours at their "main" job were more likely to have

internships. Gatta, Mary. 2023. "The class of 2023: Inequity continues to under-pin internship participation and pay status." National Association of Colleges and Employers; Hora, Matthew, Chen, Zi, Parrott, Emily, and Her, Pa. 2020. "Problematizing college internships: Exploring issues with access, program design and developmental outcomes." *Int. J. Work-Integr. Learn.* 21(3): 235–52.

88 **practicality is weighed less:** Karmarkar, Uma R., Shiv, Baba, and Knutson, Brian. 2015. "Cost conscious? The neural and behavioral impact of price pri-macy on decision making." *J. Mark. Res.* 52(4): 467–81.

88 **likely to be more effective:** Karmarkar, Shiv, and Knutson, "Cost conscious?"

88 **the more likely we are to choose it:** Smith, Stephanie M., and Krajbich, Ian. 2019. "Gaze amplifies value in decision making." *Psychol. Sci.* 30(1): 116–28.

88 **they stood to win money:** Sheng, Feng, Ramakrishnan, Arjun, Seok, Dar-sol, Zhao, Wenjia Joyce, Thelaus, Samuel, et al. 2020. "Decomposing loss aversion from gaze allocation and pupil dilation." *Proc. Natl. Acad. Sci. USA* 117(21): 11356–63.

88 **sell at unfavorable times:** Frydman, Cary, and Rangel, Antonio. 2014. "Debias-ing the disposition effect by reducing the saliency of information about a stock's purchase price." *J. Econ. Behav. Organ.* 107(Pt B): 541–52.

89 **"Of course, everybody wants":** Ernie Grunfeld in conversation with the author, March 2022.

89 **far from his family:** Grunfeld, *By the Grace of the Game.*

89 **And he did:** Livia Grunfeld in conversation with the author, March 2022.

90 **how positive or negative we feel:** Lieberman, Matthew D., Straccia, Mark A., Meyer, Meghan L., Du, Meng, and Tan, Kevin M. 2019. "Social, self, (situa-tional), and affective processes in medial prefrontal cortex (MPFC): Causal, multivariate, and reverse inference evidence." *Neurosci. Biobehav. Rev.* 99: 311–28. Activity in some parts of the value system increases when people feel a range of different emotions, including negative ones. But parts of the ven-tral striatum and medial prefrontal cortex that are core to the value system seem to correlate specifically with how positive or negative we feel. In this way, researchers have been able to use patterns of activation within the value system to predict whether people are thinking thoughts that feel subjectively nega-tive or positive to them. Lindquist, Kristen A., Satpute, Ajay B., Wager, Tor D., Weber, Jochen, and Barrett, Lisa Feldman. 2016. "The brain basis of positive and negative affect: Evidence from a meta-analysis of the human neuroimag-ing literature." *Cereb. Cortex* 26(5): 1910–22; Lieberman, Straccia, Meyer, Du, and Tan, "Social, self, (situational), and affective processes in medial prefrontal cortex (MPFC)"; Tusche, Anita, Smallwood, Jonathan, Bernhardt, Boris C., and Singer, Tania. 2014. "Classifying the wandering mind: Revealing the affec-tive content of thoughts during task-free rest periods." *NeuroImage* 97: 107–16.

91 **value system the researchers recorded:** Winecoff, Amy, Clithero, John A., Carter, R. McKell, Bergman, Sara R., Wang, Lihong, and Huettel, Scott A. 2013. "Ventro-medial prefrontal cortex encodes emotional value." *J. Neurosci.* 33(27): 11032–39.

91 **challenging emotional experiences:** Buhle, Jason T., Silvers, Jennifer A., Wager, Tor D., Lopez, Richard, Onyemekwu, Chukwudi, et al. 2014. "Cogni-

tive reappraisal of emotion: A meta-analysis of human neuroimaging studies." *Cereb. Cortex* 24(11): 2981–90.

91 **up or down, according to our goals:** Ochsner, Kevin N., and Gross, James J. 2008. "Cognitive emotion regulation: Insights from social cognitive and affective neuroscience." *Curr. Dir. Psychol. Sci.* 17(2): 153–58.

92 **silliness rather than fear:** Gay, Ross. 2019. *The Book of Delights: Essays.* Algonquin Books.

93 **following useful health advice:** Scholz, Christin, Doré, Bruce P., Cooper, Nicole, and Falk, Emily B. 2019. "Neural valuation of antidrinking campaigns and risky peer influence in daily life." *Health Psychol.* 38(7): 658–67.

93 **yourself as an objective observer:** Kross, Ethan, and Ayduk, Ozlem. 2017. "Self-distancing: Theory, research, and current directions." In *Advances in Experimental Social Psychology*, edited by James M. Olson. 55: 81–136. Elsevier Academic Press.

93 **responses to positive images:** Winecoff, Clithero, Carter, Bergman, Wang, and Huettel, "Ventromedial prefrontal cortex encodes emotional value."

94 **improve our decisions:** Kross, Ethan, and Grossmann, Igor. 2012. "Boosting wisdom: Distance from the self enhances wise reasoning, attitudes, and behavior." *J. Exp. Psychol. Gen.* 141(1): 43–48.

94 **notice how we feel:** Langer, Ellen J. 2014. *Mindfulness.* Hachette Books. 25th anniversary ed.; Ludwig, Vera U., Brown, Kirk Warren, and Brewer, Judson A. 2020. "Self-regulation without force: Can awareness leverage reward to drive behavior change?" *Perspect. Psychol. Sci.* 15(6): 1382–99.

94 **attend to smoking-related images:** Westbrook, Cecilia, Creswell, John David, Tabibnia, Golnaz, Julson, Erica, Kober, Hedy, and Tindle, Hilary A. 2013. "Mindful attention reduces neural and self-reported cue-induced craving in smokers." *Soc. Cogn. Affect. Neurosci.* 8(1): 73–84.

94 **seventeen weeks after the training:** Brewer, Judson A., Mallik, Sarah, Babuscio, Theresa A., Nich, Charla, Johnson, Hayley E., et al. 2011. "Mindfulness training for smoking cessation: Results from a randomized controlled trial." *Drug Alcohol Depend.* 119(1–2): 72–80.

94 **response to smoking-related stimuli:** Kober, Mende-Siedlecki, Kross, Weber, Mischel, et al., "Prefrontal-striatal pathway underlies cognitive regulation of craving."

94 **brain regions' reactivity to stress:** Kober, Hedy, Brewer, Judson A., Height, Keri L., and Sinha, Rajita. 2017. "Neural stress reactivity relates to smoking outcomes and differentiates between mindfulness and cognitive-behavioral treatments." *NeuroImage* 151: 4–13.

## Chapter 5: Become the Least Defensive Person You Know

98 **quality of interactions between people:** Barrick, Elyssa M., Barasch, Alixandra, and Tamir, Diana I. 2022. "The unexpected social consequences of diverting attention to our phones." *J. Exp. Soc. Psychol.* 101: 104344.

99 **wired to conflate self:** Chavez, Robert S., Heatherton, Todd F., and Wagner,

Dylan D. 2017. "Neural population decoding reveals the intrinsic positivity of the self." *Cereb. Cortex* 27(11): 5222–29.

99 **protect our self-esteem:** Taylor, Shelley E. 1989. *Positive Illusions: Creative Self-Deception and the Healthy Mind.* Basic Books.

99 **we could benefit from:** Cohen, Geoffrey L., and Sherman, David K. 2014. "The psychology of change: Self-affirmation and social psychological intervention." *Annu. Rev. Psychol.* 65: 333–71.

99 **tried to convince a loved one:** I discuss this more in a TedX Talk, "How the brain changes its mind," from which part of this paragraph was adapted.

99 **defensive when they're challenged:** Cohen and Sherman, "The psychology of change"; Hart, William, Albarracín, Dolores, Eagly, Alice H., Brechan, Inge, Lindberg, Matthew J., and Merrill, Lisa. 2009. "Feeling validated versus being correct: A meta-analysis of selective exposure to information." *Psychol. Bull.* 135(4): 555–88.

100 **"key your car":** Perih, Larysa. 2023. "82 bits of funny bad advice you should absolutely not take too seriously." Bored Panda.

101 **effectively give up money to keep it:** Kahneman, Daniel, Knetsch, Jack L., and Thaler, Richard H. 1990. "Experimental tests of the endowment effect and the Coase theorem." *J. Polit. Econ.* 98(6): 1325–48.

102 **sense of ownership attached to it:** Kahneman, Knetsch, and Thaler, "Experimental tests of the endowment effect and the Coase theorem."

103 **object and cash to begin with:** Knutson, Brian, Wimmer, G. Elliott, Rick, Scott, Hollon, Nick G., Prelec, Drazen, and Loewenstein, George. 2008. "Neural antecedents of the endowment effect." *Neuron* 58(5): 814–22.

106 **more "peripheral" ones:** Elder, Jacob, Cheung, Bernice, Davis, Tyler, and Hughes, Brent. 2023. "Mapping the self: A network approach for understanding psychological and neural representations of self-concept structure." *J. Pers. Soc. Psychol.* 124(2): 237–63.

106 **more peripheral than core:** Elder, Jacob J., Davis, Tyler H., and Hughes, Brent L. 2023. "A fluid self-concept: How the brain maintains coherence and positivity across an interconnected self-concept while incorporating social feedback." *J. Neurosci.* 43(22): 4110–28.

106 **feedback they perceived as negative:** Elder, Davis, and Hughes, "A fluid self-concept."

107 **when bad things happen:** Greenwald, Anthony G. 1980. "The totalitarian ego: Fabrication and revision of personal history." *Am. Psychol.* 35(7): 603–18.

108 **your values as *values affirmation*:** Cohen and Sherman, "The psychology of change"; Steele, Claude M. 1988. "The psychology of self-affirmation: Sustaining the integrity of the self." In *Advances in Experimental Social Psychology,* edited by Leonard Berkowitz. 21: 261–302. Academic Press.

108 **adopting new ideas and behaviors:** Falk, Emily B., O'Donnell, Matthew Brook, Cascio, Christopher N., Tinney, Francis, Kang, Yoona, et al. 2015. "Self-affirmation alters the brain's response to health messages and subsequent behavior change." *Proc. Natl. Acad. Sci. USA* 112(7): 1977–82; Cascio, Christopher N., O'Donnell, Matthew Brook, Tinney, Francis J., Lieberman,

Matthew D., Taylor, Shelley E., et al. 2016. "Self-affirmation activates brain systems associated with self-related processing and reward and is reinforced by future orientation." *Soc. Cogn. Affect. Neurosci.* 11(4): 621–29; Kang, Yoona, Cooper, Nicole, Pandey, Prateekshit, Scholz, Christin, O'Donnell, Matthew Brook, et al. 2018. "Effects of self-transcendence on neural responses to persuasive messages and health behavior change." *Proc. Natl. Acad. Sci. USA* 115(40): 9974–79.

108 **spent a lot of time sitting:** Kang, Cooper, Pandey, Scholz, O'Donnell, et al., "Effects of self-transcendence on neural responses to persuasive messages and health behavior change."

109 **decrease in their daily exercise:** Kang, Cooper, Pandey, Scholz, O'Donnell, et al., "Effects of self-transcendence on neural responses to persuasive messages and health behavior change"; Falk, O'Donnell, Cascio, Tinney, Kang, et al., "Self-affirmation alters the brain's response to health messages and subsequent behavior change."

109 **150 minutes of exercise a week:** Centers for Disease Control and Prevention. 2023. "Physical Activity for Adults: An Overview."

109 **someone with a different ideology:** Cohen, Geoffrey L., Aronson, Joshua, and Steele, Claude M. 2000. "When beliefs yield to evidence: Reducing biased evaluation by affirming the self." *Pers. Soc. Psychol. Bull.* 26(9): 1151–64.

109 **information on the harm of smoking:** Crocker, Jennifer, Niiya, Yu, and Mischkowski, Dominik. 2008. "Why does writing about important values reduce defensiveness? Self-affirmation and the role of positive other-directed feelings." *Psychol. Sci.* 19(7): 740–47.

109 **links between cancer and alcohol:** Harris, Peter R., and Napper, Lucy. 2005. "Self-affirmation and the biased processing of threatening health-risk information." *Pers. Soc. Psychol. Bull.* 31(9): 1250–63.

109 **structural racism more clearly:** Unzueta, Miguel M., and Lowery, Brian S. 2008. "Defining racism safely: The role of self-image maintenance on white Americans' conceptions of racism." *J. Exp. Soc. Psychol.* 44(6): 1491–97.

110 **parts of themselves they like:** Steele, Claude M., and Liu, Thomas J. 1983. "Dissonance processes as self-affirmation." *J. Pers. Soc. Psychol.* 45(1): 5–19.

111 **and complete each sentence:** Schimel, Jeff, Arndt, Jamie, Banko, Katherine M., and Cook, Alison. 2004. "Not all self-affirmations were created equal: The cognitive and social benefit of affirming the intrinsic (vs extrinsic) self." *Soc. Cogn.* 22(1): 75–99.

111 **growth can feel rewarding:** Walton, Gregory M., and Wilson, Timothy D. 2018. "Wise interventions: Psychological remedies for social and personal problems." *Psychol. Rev.* 125(5): 617–55.

111 **potential threats:** McQueen, Amy, and Klein, William M. P. 2006. "Experimental manipulations of self-affirmation: A systematic review." *Self Identity* 5(4): 289–354.

111 **too late for affirmation to work:** Critcher, Clayton R., Dunning, David, and Armor, David A. 2010. "When self-affirmations reduce defensiveness: Timing is key." *Pers. Soc. Psychol. Bull.* 36(7): 947–59.

111 **"like an engineered coincidence"**: Cohen and Sherman, "The psychology of change."

112 **grades in school**: Wu, Zezhen, Spreckelsen, Thees F., and Cohen, Geoffrey L. 2021. "A meta-analysis of the effect of values affirmation on academic achievement." *J. Soc. Issues* 77(3): 702–50.

112 **even if they performed well**: Cook, Jonathan E., Purdie-Vaughns, Valerie, Garcia, Julio, and Cohen, Geoffrey L. 2012. "Chronic threat and contingent belonging: Protective benefits of values affirmation on identity development." *J. Pers. Soc. Psychol.* 102(3): 479–96.

112 **advantages that accumulate**: Cohen and Sherman, "The psychology of change."

112 **difficult to later overcome**: Cook, Purdie-Vaughns, Garcia, and Cohen, "Chronic threat and contingent belonging."

113 **report feeling less purposeful**: Lewis, Nathan A., Turiano, Nicholas A., Payne, Brennan R., and Hill, Patrick L. 2017. "Purpose in life and cognitive functioning in adulthood." *Aging Neuropsychol. Cogn.* 24(6): 662–71; Cohen, Randy, Bavishi, Chirag, and Rozanski, Alan. 2016. "Purpose in life and its relationship to all-cause mortality and cardiovascular events: A meta-analysis." *Psychosom. Med.* 78(2): 122–33; Kim, Eric S., Sun, Jennifer K., Park, Nansook, and Peterson, Christopher. 2013. "Purpose in life and reduced incidence of stroke in older adults: 'The Health and Retirement Study.'" *J. Psychosom. Res.* 74(5): 427–32; Hooker, Stephanie A., and Masters, Kevin S. 2016. "Purpose in life is associated with physical activity measured by accelerometer." *J. Health Psychol.* 21(6): 962–71; Roepke, Ann Marie, Jayawickreme, Eranda, and Riffle, Olivia M. 2014. "Meaning and health: A systematic review." *Appl. Res. Qual. Life* 9(4): 1055–79; Ryff, Carol D. 2014. "Psychological well-being revisited: Advances in the science and practice of eudaimonia." *Psychother. Psychosom.* 83(1): 10–28.

113 **to take care of themselves**: Older adults who feel more purposeful tend to get more exercise and sleep and take advantage of more routine preventive health care than those with a lower sense of purpose, and broader correlations between purpose and health hold across a range of other demographic groups as well. Kim, Eric S., Shiba, Koichiro, Boehm, Julia K., and Kubzansky, Laura D. 2020. "Sense of purpose in life and five health behaviors in older adults." *Prev. Med.* 139: 106172; Kim, Eric S., Strecher, Victor J., and Ryff, Carol D. 2014. "Purpose in life and use of preventive health care services." *Proc. Natl. Acad. Sci. USA* 111(46): 16331–36; Steptoe, Andrew. 2019. "Happiness and health." *Annu. Rev. Public Health* 40: 339–59.

113 **going for a walk**: McGowan, Amanda L., Boyd, Zachary M., Kang, Yoona, Bennett, Logan, Mucha, Peter J., et al. 2023. "Within-person temporal associations among self-reported physical activity, sleep, and well-being in college students." *Psychosom. Med.* 85(2): 141–53.

113 **other constructive advice as helpful**: Kang, Yoona, Strecher, Victor J., Kim, Eric, and Falk, Emily B. 2019. "Purpose in life and conflict-related neural responses during health decision-making." *Health Psychol.* 38(6): 545–52.

114 **sense of threat in the scanner**: Kang, Cooper, Pandey, Scholz, O'Donnell, et

al., "Effects of self-transcendence on neural responses to persuasive messages and health behavior change."

114 **With a Little Help from My Friends:** (Beatles song title) Lennon–McCartney. 1967. Parlophone.

115 **changing our thoughts and actions:** Green, Melanie C., Strange, Jeffrey J., and Brock, Timothy C. 2003. *Narrative Impact: Social and Cognitive Foundations.* Taylor & Francis.

115 **tapping into the brain's social relevance system:** Mar, Raymond A. 2011. "The neural bases of social cognition and story comprehension." *Annu. Rev. Psychol.* 62: 103–34.

115 **messages or attempts to change our behavior:** Slater, Michael D., and Rouner, Donna. 2002. "Entertainment-education and elaboration likelihood: Understanding the processing of narrative persuasion." *Commun. Theory* 12(2): 173–91; Green, Melanie C., and Brock, Timothy C. 2000. "The role of transportation in the persuasiveness of public narratives." *J. Pers. Soc. Psychol.* 79(5): 701–21; de Graaf, Anneke, Hoeken, Hans, Sanders, José, and Beentjes, Johannes W. J. 2012. "Identification as a mechanism of narrative persuasion." *Communic. Res.* 39(6): 802–23.

116 **actually change the way I see myself:** Meyer, Meghan L., Zhao, Zidong, and Tamir, Diana I. 2019. "Simulating other people changes the self." *J. Exp. Psychol. Gen.* 148(11): 1898–913.

116 **participated in the study before them:** Gilead, Michael, Boccagno, Chelsea, Silverman, Melanie, Hassin, Ran R., Weber, Jochen, and Ochsner, Kevin N. 2016. "Self-regulation via neural simulation." *Proc. Natl. Acad. Sci. USA* 113(36): 10037–42.

118 **interdisciplinary collaboration that I led:** The team included my lab, Dani Bassett's lab, and David Lydon-Staley's lab at Penn; and Kevin Ochsner's lab at Columbia, Peter Mucha's lab at Dartmouth, and Zach Boyd's lab at Brigham Young University. The analysis described above was led by Mia Jovanova in my lab and made possible by research direction from Yoona Kang.

118 **peer who drank less than they did:** Jovanova, Mia, Cosme, Danielle, Doré, Bruce, Kang, Yoona, Stanoi, Ovidia, et al. 2023. "Psychological distance intervention reminders reduce alcohol consumption frequency in daily life." *Sci. Rep.* 13(1): 12045.

119 **when they reacted naturally:** Although students' self-reports of their drinking aren't perfect, there's no reason to believe that they would be systematically better or worse at reporting in one condition or the other, so their reports provide a good indication of the intervention's effectiveness.

120 **hometown team, the Cleveland Cavaliers:** Gainsburg, Izzy, Sowden, Walter J., Drake, Brittany, Herold, Warren, and Kross, Ethan. 2022. "Distanced self-talk increases rational self-interest." *Sci. Rep.* 12(1): 511; Kross, Ethan. 2021. *Chatter: The Voice in Our Head, Why It Matters, and How to Harness It.* Crown.

120 **"One thing I didn't want":** xNeonHero. 2010. *LeBron James Talks about LeBron James.* https://www.youtube.com/watch?v=yrd9T-hny84.

120 **third-person perspective:** St. Jacques, Peggy L., Szpunar, Karl K., and Schacter, Daniel L. 2017. "Shifting visual perspective during retrieval shapes

autobiographical memories." *NeuroImage* 148: 103–14; Mischkowski, Dominik, Kross, Ethan, and Bushman, Brad J. 2012. "Flies on the wall are less aggressive: Self-distancing 'in the heat of the moment' reduces aggressive thoughts, angry feelings and aggressive behavior." *J. Exp. Soc. Psychol.* 48(5): 1187–91; Kross, Ethan, Duckworth, Angela, Ayduk, Ozlem, Tsukayama, Eli, and Mischel, Walter. 2011. "The effect of self-distancing on adaptive versus maladaptive self-reflection in children." *Emotion* 11(5): 1032–39; Ayduk, Ozlem, and Kross, Ethan. 2010. "From a distance: Implications of spontaneous self-distancing for adaptive self-reflection." *J. Pers. Soc. Psychol.* 98(5): 809–29.

120 **our ability to solve problems:** Kross, Duckworth, Ayduk, Tsukayama, and Mischel, "The effect of self-distancing on adaptive versus maladaptive self-reflection in children"; Ayduk and Kross, "From a distance."

120 **a "fly on the wall":** Mischkowski, Kross, and Bushman, "Flies on the wall are less aggressive."

120 **reduces aggressive thoughts and reactions:** Likewise, telling our stories as if they happened to someone else or considering how we would advise another person about a problem can give us new perspective, make negative emotions feel less intense, help us come up with wiser solutions, and change our own behavior. Grossmann, Igor, and Kross, Ethan. 2014. "Exploring Solomon's paradox: Self-distancing eliminates the self-other asymmetry in wise reasoning about close relationships in younger and older adults." *Psychol. Sci.* 25(8): 1571–80; Eskreis-Winkler, Lauren, Milkman, Katherine L., Gromet, Dena M., and Duckworth, Angela L. 2019. "A large-scale field experiment shows giving advice improves academic outcomes for the advisor." *Proc. Natl. Acad. Sci. USA* 116(30): 14808–10; St. Jacques, Szpunar, and Schacter, "Shifting visual perspective during retrieval shapes autobiographical memories."

120 **changes how the brain works:** Buhle, Jason T., Silvers, Jennifer A., Wager, Tor D., Lopez, Richard, Onyemekwu, Chukwudi, et al. 2014. "Cognitive reappraisal of emotion: A meta-analysis of human neuroimaging studies." *Cereb. Cortex* 24(11): 2981–90.

121 **they reacted as they naturally would:** Zhou, Dale, Kang, Yoona, Cosme, Danielle, Jovanova, Mia, He, Xiaosong, et al. 2023. "Mindful attention promotes control of brain network dynamics for self-regulation and discontinues the past from the present." *Proc. Natl. Acad. Sci. USA* 120(2): e2201074119.

121 **move on to the next:** Zhou, Kang, Cosme, Jovanova, He, et al., "Mindful attention promotes control of brain network dynamics for self-regulation and discontinues the past from the present."

121 **forms of meditation and mindfulness:** Kang, Yoona, Gruber, June, and Gray, Jeremy R. 2013. "Mindfulness and de-automatization." *Emot. Rev.* 5(2): 192–201; Ie, Amanda, Ngnoumen, Christelle T., and Langer, Ellen J. 2014. *The Wiley Blackwell Handbook of Mindfulness.* John Wiley & Sons; Langer, *Mindfulness.*

121 **automatic responses to everyday events:** Tang, Yi-Yuan, Hölzel, Britta K., and Posner, Michael I. 2015. "The neuroscience of mindfulness meditation." *Nat. Rev. Neurosci.* 16(4): 213–25.

121 **in response to distressing images:** Kober, Hedy, Buhle, Jason, Weber, Jochen, Ochsner, Kevin N., and Wager, Tor D. 2019. "Let it be: Mindful acceptance down-regulates pain and negative emotion." *Soc. Cogn. Affect. Neurosci.* 14(11): 1147–58.

121 **cravings for alcohol:** Suzuki, Shosuke, Mell, Maggie Mae, O'Malley, Stephanie S., Krystal, John H., Anticevic, Alan, and Kober, Hedy. 2020. "Regulation of craving and negative emotion in alcohol use disorder." *Biol. Psychiatry Cogn. Neurosci. Neuroimaging* 5(2): 239–50; Naqvi, Nasir H., Ochsner, Kevin N., Kober, Hedy, Kuerbis, Alexis, Feng, Tianshu, et al. 2015. "Cognitive regulation of craving in alcohol-dependent and social drinkers." *Alcohol. Clin. Exp. Res.* 39(2): 343–49.

121 **cigarettes:** Brewer, Judson A., Mallik, Sarah, Babuscio, Theresa A., Nich, Charla, Johnson, Hayley E., et al. 2011. "Mindfulness training for smoking cessation: Results from a randomized controlled trial." *Drug Alcohol Depend.* 119(1–2): 72–80; Westbrook, Cecilia, Creswell, John David, Tabibnia, Golnaz, Julson, Erica, Kober, Hedy, and Tindle, Hilary A. 2013. "Mindful attention reduces neural and self-reported cue-induced craving in smokers." *Soc. Cogn. Affect. Neurosci.* 8(1): 73–84.

121 **unhealthy foods:** Sun, Wendy, and Kober, Hedy. 2020. "Regulating food craving: From mechanisms to interventions." *Physiol. Behav.* 222: 112878.

121 **and other drugs:** DeVito, Elise E., Worhunsky, Patrick D., Carroll, Kathleen M., Rounsaville, Bruce J., Kober, Hedy, and Potenza, Marc N. 2012. "A preliminary study of the neural effects of behavioral therapy for substance use disorders." *Drug Alcohol Depend.* 122(3): 228–35.

122 **more nimble and present-focused:** Zhou, Kang, Cosme, Jovanova, He, et al., "Mindful attention promotes control of brain network dynamics for self-regulation and discontinues the past from the present."

## *Chapter 6:* Connect the Dots

124 **keep track of complex social information:** Dunbar, R. I. M., and Shultz, Susanne. 2007. "Evolution in the social brain." *Science* 317(5843): 1344–47.

124 **people who are most like us:** Merritt, Carrington C., MacCormack, Jennifer K., Stein, Andrea G., Lindquist, Kristen A., and Muscatell, Keely A. 2021. "The neural underpinnings of intergroup social cognition: An fMRI meta-analysis." *Soc. Cogn. Affect. Neurosci.* 16(9): 903–14.

124 **validate our existing beliefs and preferences:** Hart, William, Albarracín, Dolores, Eagly, Alice H., Brechan, Inge, Lindberg, Matthew J., and Merrill, Lisa. 2009. "Feeling validated versus being correct: A meta-analysis of selective exposure to information." *Psychol. Bull.* 135(4): 555–88.

124 **upward of five million people each week:** Young, Robin, and Tong, Scott. About. *Here and Now.* WBUR.

125 **they can leverage to solve problems:** Burt, Ronald S. 2004. "Structural holes and good ideas." *Am. J. Sociol.* 110(2): 349–99.

125 **more readily as leaders:** Burt, Ronald S., Kilduff, Martin, and Tasselli, Stefano.

2013. "Social network analysis: Foundations and frontiers on advantage." *Annu. Rev. Psychol.* 64: 527–47; Burt, Ronald S. 2005. *Brokerage and Closure: An Introduction to Social Capital.* Oxford University Press; Burt, Ronald S., Reagans, Ray E., and Volvovsky, Hagay C. 2021. "Network brokerage and the perception of leadership." *Soc. Networks* 65: 33–50.

125 **new opportunities for connection:** Kleinbaum, Adam M., and Stuart, Toby E. 2014. "Inside the black box of the corporate staff: Social networks and the implementation of corporate strategy." *Strategic Manage. J.* 35(1): 24–47.

127 **he brought me on board as a consultant:** Seydel, Roland (Innovation Explorer at Adidas), in personal discussion with the author, November 2023.

127 **solve problems creatively and productively:** Burt, "Structural holes and good ideas."

128 **building materials or rock formations:** Tompson, Steven H., Kahn, Ari E., Falk, Emily B., Vettel, Jean M., and Bassett, Danielle S. 2020. "Functional brain network architecture supporting the learning of social networks in humans." *NeuroImage* 210: 116498.

128 **in a social network, and vice versa:** Tompson, Steven H., Kahn, Ari E., Falk, Emily B., Vettel, Jean M., and Bassett, Danielle S. 2019. "Individual differences in learning social and nonsocial network structures." *J. Exp. Psychol. Learn. Mem. Cogn.* 45(2): 253–71.

128 **one path to building useful bridges:** Zhang, Aven, and Kleinbaum, "License to broker"; Kleinbaum, Adam M. 2012. "Organizational misfits and the origins of brokerage in intrafirm networks." *Adm. Sci. Q.* 57(3): 407–52.

129 **tend to be seen as leaders:** Burt, Reagans, and Volvovsky, "Network brokerage and the perception of leadership."

129 **active participants in the training:** Burt, Ronald S., and Ronchi, Don. 2007. "Teaching executives to see social capital: Results from a field experiment." *Soc. Sci. Res.* 36(3): 1156–83.

129 **thinking about new programs:** Baer, M., Evans, K., Oldham, G. R., and Boasso, A. 2015. "The social network side of individual innovation." *Org. Psychol. Rev.* 5(3): 191–223.

129 **"You could see people we wanted":** Hambach, Philip (Director of Global Consumer Insights at Adidas), in discussion with the author, November 2023.

131 **others in our networks:** Zerubavel, Noam, Bearman, Peter S., Weber, Jochen, and Ochsner, Kevin N. 2015. "Neural mechanisms tracking popularity in real-world social networks." *Proc. Natl. Acad. Sci. USA* 112(49): 15072–77; Morelli, Sylvia A., Leong, Yuan Chang, Carlson, Ryan W., Kullar, Monica, and Zaki, Jamil. 2018. "Neural detection of socially valued community members." *Proc. Natl. Acad. Sci. USA* 115(32): 8149–54.

132 **nature of those relationships:** Parkinson, Carolyn, Kleinbaum, Adam M., and Wheatley, Thalia. 2017. "Spontaneous neural encoding of social network position." *Nat. Hum. Behav.* 1(5): 0072.

132 **popular and well connected others are:** Zerubavel, Bearman, Weber, and Ochsner, "Neural mechanisms tracking popularity in real-world social networks."

132 **social support and empathy to others:** Morelli, Leong, Carlson, Kullar, and Zaki, "Neural detection of socially valued community members."

132 **many others in their social networks:** Morelli, Leong, Carlson, Kullar, and Zaki, "Neural detection of socially valued community members."

133 **have valuable perspectives too:** Cikara, Mina, and Van Bavel, Jay J. 2014. "The neuroscience of intergroup relations: An integrative review." *Perspect. Psychol. Sci.* 9(3): 245–74.

133 **shine light on new possibilities:** Antonio, Anthony Lising, Chang, Mitchell J., Hakuta, Kenji, Kenny, David A., Levin, Shana, and Milem, Jeffrey F. 2004. "Effects of racial diversity on complex thinking in college students." *Psychol. Sci.* 15(8): 507–10; Díaz-García, Cristina, González-Moreno, Angela, and Jose Sáez-Martínez, Francisco. 2013. "Gender diversity within R&D teams: Its impact on radicalness of innovation." *Innovations* 15(2): 149–60.

134 **part of their out-group:** Merritt, MacCormack, Stein, Lindquist, and Muscatell, "The neural underpinnings of intergroup social cognition: An fMRI meta-analysis."

134 **their collective team performance:** Poleacovschi, Cristina, Faust, Kasey, Roy, Arkajyoti, and Feinstein, Scott. 2021. "Identity of engineering expertise: Implicitly biased and sustaining the gender gap." *J. Civ. Eng. Educ.* 147(1): 04020011; Neal, Tess M. S. 2014. "Women as expert witnesses: A review of the literature." *Behav. Sci. Law* 32(2): 164–79; Nelson, Larry R., Signorella, Margaret L., and Botti, Karin G. 2016. "Accent, gender, and perceived competence." *Hisp. J. Behav. Sci.* 38(2): 166–85; Baugh, S. Gayle, and Graen, George B. 1997. "Effects of team gender and racial composition on perceptions of team performance in cross-functional teams." *Group Organ. Manag.* 22(3): 366–83.

134 **tend to be more novel:** Hofstra, Bas, Kulkarni, Vivek V., Galvez, Sebastian Munoz Najar, He, Bryan, Jurafsky, Dan, and McFarland, Daniel A. 2020. "The diversity–innovation paradox in science." *Proc. Natl. Acad. Sci. USA* 117(17): 9284–91.

134 **more generative in their fields:** Chilton, Adam, Driver, Justin, Masur, Jonathan S., and Rozema, Kyle. 2022. "Assessing affirmative action's diversity rationale." *Columbia Law Rev.* 122(2): 331–406.

135 **the pursuit of social justice:** brown, adrienne maree. 2019. *Pleasure Activism*. AK Press.

135 **we began to explore earlier:** Burt, "Structural holes and good ideas"; Hargadon, Andrew. 2013. "Brokerage and innovation." In *Oxford Handbook of Innovation Management*, edited by Mark Dodgson, David M. Gann, and Nelson Phillips, 163–80. Oxford University Press.

135 **different social circles:** Burt, Ronald S., and Soda, Giuseppe. 2017. "Social origins of great strategies." *Strategy Sci.* 2(4): 226–33.

135 **more diverse knowledge and expertise:** Goldberg, Amir, Srivastava, Sameer B., Manian, V. Govind, Monroe, William, and Potts, Christopher. 2016. "Fitting in or standing out? The tradeoffs of structural and cultural embeddedness." *Am. Sociol. Rev.* 81(6): 1190–222.

136 **In the face of this problem:** Dworkin, Jordan D., Linn, Kristin A., Teich, Erin G., Zurn, Perry, Shinohara, Russell T., and Bassett, Danielle S. 2020. "The extent and drivers of gender imbalance in neuroscience reference lists." *Nat. Neurosci.* 23(8): 918–26; Bertolero, Maxwell A., Dworkin, Jordan D., David, Sophia U., López Lloreda, Claudia López, Srivastava, Pragya, et al. 2020. "Racial and ethnic imbalance in neuroscience reference lists and intersections with gender." Working paper. Preprint doi:10.1101/2020.10.12.336230; Wang, Xinyi, Dworkin, Jordan D., Zhou, Dale, Stiso, Jennifer, Falk, Emily B., et al. 2021. "Gendered citation practices in the field of communication." *Ann. Int. Commun. Assoc.* 45(2): 134–53; Chakravartty, Paula, Kuo, Rachel, Grubbs, Victoria, and McIlwain, Charlton. 2018. "#CommunicationSoWhite." *J. Commun.* 68(2): 254–66; Caplar, Neven, Tacchella, Sandro, and Birrer, Simon. 2017. "Quantitative evaluation of gender bias in astronomical publications from citation counts." *Nat. Astron.* 1(6): 0141; Dion, Michelle L., Sumner, Jane Lawrence, and Mitchell, Sara McLaughlin. 2018. "Gendered citation patterns across political science and social science methodology fields." *Polit. Anal.* 26(3): 312–27; Maliniak, Daniel, Powers, Ryan, and Walter, Barbara F. 2013. "The gender citation gap in international relations." *Int. Organ.* 67(4): 889–922; Fulvio, Jacqueline M., Akinnola, Ileri, and Postle, Bradley R. 2021. "Gender (im)balance in citation practices in cognitive neuroscience." *J. Cogn. Neurosci.* 33(1): 3–7.

136 **tools for academics:** Zurn, Perry, Bassett, Danielle S., and Rust, Nicole C. 2020. "The citation diversity statement: A practice of transparency, a way of life." *Trends Cogn. Sci.* 24(9): 669–72.

137 **paying attention to the views of others:** Galinsky, Adam, and Schweitzer, Maurice. 2016. "Why every great leader needs to be a great perspective taker." *Leader to Leader* 2016(80): 32–37; Galinsky, Adam D., Magee, Joe C., Rus, Diana, Rothman, Naomi B., and Todd, Andrew R. 2014. "Acceleration with steering: The synergistic benefits of combining power and perspective-taking." *Soc. Psychol. Personal. Sci.* 5(6): 627–35.

137 **to connect with others actually grow:** Burt and Ronchi, "Teaching executives to see social capital"; Kwon, Seok Woo, Rondi, Emanuela, Levin, Daniel Z., De Massis, Alfredo, and Brass, Daniel J. 2020. "Network brokerage: An integrative review and future research agenda." *J. Manage.* 46(6): 1092–120; Burt, Reagans, and Volvovsky, "Network brokerage and the perception of leadership."

137 **making the effort to connect:** Kwon, Rondi, Levin, De Massis, and Brass, "Network brokerage."

138 **a key leadership skill:** Galinsky and Schweitzer, "Why every great leader needs to be a great perspective taker."

138 **reach better deals:** Galinsky, Adam D., Maddux, William W., Gilin, Debra, and White, Judith B. 2008. "Why it pays to get inside the head of your opponent: The differential effects of perspective taking and empathy in negotiations." *Psychol. Sci.* 19(4) :378–384; Galinsky and Schweitzer, "Why every great leader needs to be a great perspective taker."

138 **create value for both parties:** Galinsky, Maddux, Gilin, and White, "Why it pays to get inside the head of your opponent."

138 **other people's thoughts and feelings:** Muscatell, Keely A., Morelli, Sylvia A., Falk, Emily B., Way, Baldwin M., Pfeifer, Jennifer H., et al. 2012. "Social status modulates neural activity in the mentalizing network." *NeuroImage* 60(3): 1771–77.

138 **faces showing negative emotions:** Muscatell, Morelli, Falk, Way, Pfeifer, et al., "Social status modulates neural activity in the mentalizing network."

138 **an essential leadership skill:** Galinsky and Schweitzer, "Why every great leader needs to be a great perspective taker"; Galinsky, Magee, Rus, Rothman, and Todd, "Acceleration with steering."

138 **social relevance system online:** Hildebrandt, Malin K., Jauk, Emanuel, Lehmann, Konrad, Maliske, Lara, and Kanske, Philipp. 2021. "Brain activation during social cognition predicts everyday perspective-taking: A combined fMRI and ecological momentary assessment study of the social brain." *NeuroImage* 227: 117624.

139 **NYU, the University of Groningen:** Galinsky, Magee, Rus, Rothman, and Todd, "Acceleration with steering."

139 **identified the suspect more often:** Galinsky, Magee, Rus, Rothman, and Todd, "Acceleration with steering."

140 **often we are wrong:** Eyal, Tal, Steffel, Mary, and Epley, Nicholas. 2018. "Perspective mistaking: Accurately understanding the mind of another requires getting perspective, not taking perspective." *J. Pers. Soc. Psychol.* 114(4): 547–71.

140 **therefore what they will do:** Thornton, Mark A., and Tamir, Diana I. 2021. "People accurately predict the transition probabilities between actions." *Sci. Adv.* 7(9): eabd4995.

140 **find out for sure:** Eyal, Steffel, and Epley, "Perspective mistaking."

140 **improve our relationships with them:** Bruneau, Emile G., and Saxe, Rebecca. 2012. "The power of being heard: The benefits of 'perspective-giving' in the context of intergroup conflict." *J. Exp. Soc. Psychol.* 48(4): 855–66.

140 **as does the research:** Fang, Ruolian, Landis, Blaine, Zhang, Zhen, Anderson, Marc H., Shaw, Jason D., and Kilduff, Martin. 2015. "Integrating personality and social networks: A meta-analysis of personality, network position, and work outcomes in organizations." *Organ. Sci.* 26(4): 1243–60.

141 **knowledge, and ways of knowing:** Zurn, Perry, and Bassett, Dani S. 2022. *Curious Minds: The Power of Connection.* MIT Press.

141 **people might connect to one another:** Zurn, and Bassett, *Curious Minds.*

## *Chapter 7*: Getting in Sync

146 **teens at a high school dance:** Denworth, Lydia. 2023. "Brain waves synchronize when people interact." *Scientific American*, July 1, 2023, 50.

146 **behavior of pairs of bats:** Zhang, Wujie, and Yartsev, Michael M. 2019. "Correlated neural activity across the brains of socially interacting bats." *Cell* 178(2): 413–28.

147 **one person's brain signals:** Burns, Shannon M., Tsoi, Lily, Falk, Emily B., Speer, Sebastian P. H., Mwilambwe-Tshilobo, Laetitia, and Tamir, Diana I. Forthcoming. "Interdependent minds: Quantifying the dynamics of successful social interaction." *Current Directions in Psychological Science.* Research highlights several different kinds of interdependence in people's brain responses during communication. Burns et al. highlight examples, including synchrony, where two people's brains do the same thing at the same time; meta-stable synchrony, where the degree of alignment waxes and wanes over time; recurrence, where people's brains follow the same pattern but at a temporal lag—the first person might lead and then the second person might mirror what the first person's brain did; and complementarity, where two brains are coordinated in their activation, but doing different things.

147 **Happy Together:** (Turtles song title) Bonner, Garry, and Gordon, Alan. 1967. White Whale.

148 **easy to get in sync with others:** Kokal, Idil, Engel, Annerose, Kirschner, Sebastian, and Keysers, Christian. 2011. "Synchronized drumming enhances activity in the caudate and facilitates prosocial commitment—if the rhythm comes easily." *PLoS One* 6(11): e27272.

148 **system and not a real person:** Cacioppo, S., Zhou, H., Monteleone, G., Majka, E. A., Quinn, K. A., et al. 2014. "You are in sync with me: Neural correlates of interpersonal synchrony with a partner." *Neuroscience* 277: 842–58.

148 **unpleasant experiences with others:** Peng, Weiwei, Lou, Wutao, Huang, Xiaoxuan, Ye, Qian, Tong, Raymond Kai Yu, and Cui, Fang. 2021. "Suffer together, bond together: Brain-to-brain synchronization and mutual affective empathy when sharing painful experiences." *NeuroImage* 238: 118249.

148 **when solving it alone:** Krill, Austen L., and Platek, Steven M. 2012. "Working together may be better: Activation of reward centers during a cooperative maze task." *PLoS One* 7(2): e30613.

148 **when this alignment occurs:** Shamay-Tsoory, Simone G., Saporta, Nira, Marton-Alper, Inbar Z., and Gvirts, Hila Z. 2019. "Herding brains: A core neural mechanism for social alignment." *Trends Cogn. Sci.* 23(3): 174–86.

149 **had happened in the story later:** Stephens, Greg J., Silbert, Lauren J., and Hasson, Uri. 2010. "Speaker–listener neural coupling underlies successful communication." *Proc. Natl. Acad. Sci. USA* 107(32): 14425–30.

149 **facts from the teacher's lecture:** Nguyen, Mai, Chang, Ashley, Micciche, Emily, Meshulam, Meir, Nastase, Samuel A., and Hasson, Uri. 2021. "Teacher–student neural coupling during teaching and learning." *Soc. Cogn. Affect. Neurosci.* 17(4): 367–76.

149 **the better they performed:** Reinero, Diego A., Dikker, Suzanne, and Van Bavel, Jay J. 2021. "Inter-brain synchrony in teams predicts collective performance." *Soc. Cogn. Affect. Neurosci.* 16(1–2): 43–57.

149 **someone else might not have had yet:** Zadbood, Asieh, Chen, Janice, Leong, Yuan Chang, Norman, Kenneth A., and Hasson, Uri. 2017. "How we transmit memories to other brains: Constructing shared neural representations via communication." *Cereb. Cortex* 27(10): 4988–5000.

150 **humor from *America's Funniest Home Videos*:** Parkinson, Carolyn, Kleinbaum, Adam M., and Wheatley, Thalia. 2018. "Similar neural responses predict friendship." *Nat. Commun.* 9(1): 332.

150 **showed less similar responses:** Parkinson, Kleinbaum, and Wheatley, "Similar neural responses predict friendship."

150 **become friends as the year progressed:** Carolyn Parkinson and Thalia Wheatley in discussion with the author, January 2024.

151 **people they aren't close to:** Hyon, Ryan, Kleinbaum, Adam M., and Parkinson, Carolyn. 2020. "Social network proximity predicts similar trajectories of psychological states: Evidence from multi-voxel spatiotemporal dynamics." *NeuroImage* 216: 116492.

151 **residents of a Korean fishing village:** Hyon, Ryan, Youm, Yoosik, Kim, Junsol, Chey, Jeanyung, Kwak, Seyul, and Parkinson, Carolyn. 2020. "Similarity in functional brain connectivity at rest predicts interpersonal closeness in the social network of an entire village." *Proc. Natl. Acad. Sci. USA* 117(52): 33149–60.

151 **compared with someone watching CNN:** Broockman, David, and Kalla, Joshua. 2022. "The impacts of selective partisan media exposure: A field experiment with Fox News viewers." Working paper. Preprint doi:10.31219/osf.io/jrw26.

151 **from violence:** Gerbner, George, and Gross, Larry. 2006. "Living with television: The violence profile." *J. Commun.* 26(2): 172–99; Gerbner, George, Gross, Larry, Morgan, Michael, and Signorielli, Nancy. 1986. "Living with television: The dynamics of the cultivation process." In *Perspectives on Media Effects*, edited by Jennings Bryant and Dolf Zillmann, 17–40. Routledge; Gerbner, George, Gross, Larry, Morgan, Michael, and Signorielli, Nancy. 1980. "The 'mainstreaming' of America: Violence profile no. 11." *J. Commun.* 30(3): 10–29.

151 **to gender:** Aubrey, Jennifer Stevens, and Harrison, Kristen. 2004. "The gender-role content of children's favorite television programs and its links to their gender-related perceptions." *Media Psychol.* 6(2): 111–46; Scharrer, Erica, and Blackburn, Greg. 2018. "Cultivating conceptions of masculinity: Television and perceptions of masculine gender role norms." *Mass Commun. Soc.* 21(2): 149–77.

151 **sexuality:** Żerebecki, Bartosz G., Opree, Suzanna J., Hofhuis, Joep, and Janssen, Susanne. 2021. "Can TV shows promote acceptance of sexual and ethnic minorities? A literature review of television effects on diversity attitudes." *Sociol. Compass* 15(8): e12906; Taylor, Laramie D. 2005. "Effects of visual and verbal sexual television content and perceived realism on attitudes and beliefs." *J. Sex Res.* 42(2): 130–37.

151 **race:** Punyanunt-Carter, Narissra M. 2008. "The perceived realism of African American portrayals on television." *Howard J. Commun.* 19(3): 241–57; Busselle, Rick, and Crandall, Heather. 2002. "Television viewing and perceptions about race differences in socioeconomic success." *J. Broadcast. Electron. Media* 46(2): 265–82.

151 **about the justice system:** Mutz, Diana C., and Nir, Lilach. 2010. "Not necessarily the news: Does fictional television influence real-world policy preferences?" *Mass Commun. Soc.* 13(2): 196–217.

151 **or immigration:** Atwell Seate, Anita, and Mastro, Dana. 2016. "Media's influ-

ence on immigration attitudes: An intergroup threat theory approach." *Commun. Monogr.* 83(2): 194–213.

152 **one study led by Uri Hasson:** Hasson, Uri, Ghazanfar, Asif A., Galantucci, Bruno, Garrod, Simon, and Keysers, Christian. 2012. "Brain-to-brain coupling: A mechanism for creating and sharing a social world." *Trends Cogn. Sci.* 16(2): 114–21.

152 **brain responses to political issues:** Broockman and Kalla, "The impacts of selective partisan media exposure"; Schmälzle, Ralf, Häcker, Frank E. K., Honey, Christopher J., and Hasson, Uri. 2015. "Engaged listeners: Shared neural processing of powerful political speeches." *Soc. Cogn. Affect. Neurosci.* 10(8): 1137–43.

152 **same media and ideas as we are:** Van Baar, Jeroen M., Halpern, David J., and FeldmanHall, Oriel. 2021. "Intolerance of uncertainty modulates brain-to-brain synchrony during politically polarized perception." *Proc. Natl. Acad. Sci. USA* 118(20): e2022491118; Jacoby, Nir, Landau-Wells, Markia, Pearl, Jacob, Paul, Alexandra, Falk, Emily B., Bruneau, Emile G., and Ochsner, Kevin N. 2024. "Partisans process policy-based and identity-based messages using dissociable neural systems." *Cereb. Cortex* 34(9): bhae368.

153 **those with different identities:** Leong, Yuan Chang, Chen, Janice, Willer, Robb, and Zaki, Jamil. 2020. "Conservative and liberal attitudes drive polarized neural responses to political content." *Proc. Natl. Acad. Sci. USA* 117(44): 27731–39.

153 **kept watching their typical news:** Broockman and Kalla, "The impacts of selective partisan media exposure."

154 **"You can't trust her":** Salinger, J. D. "Pretty Mouth and Green My Eyes." *New Yorker*, July 14, 1951.

154 **shared context biases our brains:** Yeshurun, Yaara, Swanson, Stephen, Simony, Erez, Chen, Janice, Lazaridi, Christina, et al. 2017. "Same story, different story: The neural representation of interpretive frameworks." *Psychol. Sci.* 28(3): 307–19.

154 **Arthur is simply paranoid:** If you are curious, here are the full backstories the researchers gave to the two groups of volunteers. The first, in which the woman in Lee's bed is Arthur's wife, Joanie:

"It is late at night and the phone is ringing. On one end of the line is Arthur; Arthur just came home from a party. He left the party without finding his wife, Joanie. As always, Joanie was flirting with everybody at the party. Arthur is very upset. On the other end is Lee, Arthur's friend. He is at home with Joanie, Arthur's wife. Lee and Joanie have just returned from the same party. They have been having an affair for over a year now. They are thinking about the excuse Lee will use to calm Arthur this time."

The second, in which Lee and Joanie are not having an affair:

"It is late at night and the phone is ringing. On one end of the line is Arthur; Arthur just came home from a party. He left the party without finding his wife, Joanie. As always, Arthur is paranoid, worrying that she might be having an affair, which is not true. On the other end is Lee, Arthur's friend. He is at home with his girlfriend, Rose. Lee and Rose have just returned from the same party and are desperate to go to sleep. They do not know anything about Joanie's whereabouts and are tired of dealing with Arthur's overreactions."

154 **similar ways to other group members:** Yeshurun, Swanson, Simony, Chen, Lazaridi, et al., "Same story, different story."

155 **Maya's thoughts and actions are headed:** Thornton, Mark A., and Tamir, Diana I. 2021. "People accurately predict the transition probabilities between actions." *Sci. Adv.* 7(9): eabd4995.

156 **harmful behaviors associated with them:** Moore-Berg, Samantha L., Ankori-Karlinsky, Lee Or, Hameiri, Boaz, and Bruneau, Emile. 2020. "Exaggerated meta-perceptions predict intergroup hostility between American political partisans." *Proc. Natl. Acad. Sci. USA* 117(26): 14864–72; Lees, Jeffrey, and Cikara, Mina. 2020. "Inaccurate group meta-perceptions drive negative out-group attributions in competitive contexts." *Nat. Hum. Behav.* 4(3): 279–86; Landry, Alexander P., Schooler, Jonathan W., Willer, Robb, and Seli, Paul. 2023. "Reducing explicit blatant dehumanization by correcting exaggerated meta-perceptions." *Soc. Psychol. Personal. Sci.* 14(4): 407–18.

157 **into alignment during the conversation itself:** Mwilambwe-Tshilobo, L., Tsoi, L., Speer, S., Burns, S., Falk, E., and Tamir, D. Forthcoming. "Real-time conversation leads to neural alignment in friends and strangers."

157 **with others in a group:** Sievers, Beau, Welker, Christopher, Hasson, Uri, Kleinbaum, Adam M., Wheatley, Thalia, and Way, Jane Stanford. 2024. "Consensus-building conversation leads to neural alignment." *Nat. Commun.* 15(1), 3936.

158 **or maybe even undermine it:** Kleinbaum, Adam (Professor of Business Administration at Dartmouth), in discussion with the author, June 2024.

159 **people like fast-paced:** Templeton, Emma M., Chang, Luke J., Reynolds, Elizabeth A., Cone LeBeaumont, Marie D., and Wheatley, Thalia. 2022. "Fast response times signal social connection in conversation." *Proc. Natl. Acad. Sci. USA* 119(4): e2116915119.

159 **deep:** Kardas, Michael, Kumar, Amit, and Epley, Nicholas. 2022. "Overly shallow? Miscalibrated expectations create a barrier to deeper conversation." *J. Pers. Soc. Psychol.* 122(3): 367–98.

159 **where they learn new things:** Cooney, Gus, Gilbert, Daniel T., and Wilson, Timothy D. 2017. "The novelty penalty: Why do people like talking about new experiences but hearing about old ones?" *Psychol. Sci.* 28(3): 380–94; Westgate, Erin C., and Wilson, Timothy D. 2018. "Boring thoughts and bored minds: The MAC model of boredom and cognitive engagement." *Psychol. Rev.* 125(5): 689–713.

159 **talk about novel ideas:** See, Abigail, Roller, Stephen, Kiela, Douwe, and Weston, Jason. 2019. "What makes a good conversation? How controllable attributes affect human judgments." In *Proceedings of the 2019 Conference of the North American Chapter of the Association for Computational Linguistics: Human Language Technologies*, Vol. 1 (Long and Short Papers), edited by Jill Burstein, Christy Doran, and Thamar Solorio, 1702–23.

160 **game called "Fast Friends":** Speer, Sebastian, Mwilambwe-Tshilobo, Laetitia, Tsoi, Lily, Burns, Shannon M., Falk, Emily B., and Tamir, Diana. 2024. "What

makes a good conversation? fMRI-hyperscanning shows friends explore and strangers converge." *Nat. Commun.* 15(1): 7781.

160 **expect to like one another:** Aron, Arthur, Melinat, Edward, Aron, Elaine N., Vallone, Robert Darrin, and Bator, Renee J. 1997. "The experimental generation of interpersonal closeness: A procedure and some preliminary findings." *Pers. Soc. Psychol. Bull.* 23(4): 363–77.

161 **often have more fun:** Templeton, Emma M. 2023. "What makes conversation good? How responsivity, topics, and insider language predict feelings of connection." PhD thesis. Dartmouth College.

161 **interesting conversation partners:** See, Roller, Kiela, and Weston, "What makes a good conversation? How controllable attributes affect human judgments"; Templeton, Chang, Reynolds, Cone LeBeaumont, and Wheatley, "Fast response times signal social connection in conversation"; Kardas, Kumar, and Epley, "Overly shallow?"

161 **complement one another:** Burns, Shannon. 2020. "Neural and Psychological Coordination in Social Communication and Interaction." PhD thesis. University of California, Los Angeles; Burns, Tsoi, Falk, Speer, Mwilambwe-Tshilobo, and Tamir, "Interdependent minds: Quantifying the dynamics of successful social interaction."

162 **to solve the problems:** Speer, Sebastian, Sened, Haran, Mwilambwe-Tshilobo, Laetitia, Tsoi, Lily, Burns, Shannon M., Falk, Emily B., and Tamir, Diana I. 2024. "Finding agreement: fMRI-hyperscanning reveals that dyads diverge in mental state space to align opinions." BioRxiv Preprint

162 **increasing productive self-reflection:** Itzchakov, Guy, Weinstein, Netta, Leary, Mark, Saluk, Dvori, and Amar, Moty. 2024. "Listening to understand: The role of high-quality listening on speakers' attitude depolarization during disagreements." *J. Pers. Soc. Psychol.* 126(2): 213–39; Bruneau, Emile G., and Saxe, Rebecca. 2012. "The power of being heard: The benefits of 'perspective-giving' in the context of intergroup conflict." *J. Exp. Soc. Psychol.* 48(4): 855–66.

162 **support constructive connection:** Santos, Luiza A., Voelkel, Jan G., Willer, Robb, and Zaki, Jamil. 2022. "Belief in the utility of cross-partisan empathy reduces partisan animosity and facilitates political persuasion." *Psychol. Sci.* 33(9): 1557–73.

## *Chapter 8*: Small Acts of Sharing

167 **engaged on online platforms:** Hancock, Jeff, Liu, Sunny Xun, Luo, Mufan, and Mieczkowski, Hannah. 2022. "Psychological well-being and social media use: A meta-analysis of associations between social media use and depression, anxiety, loneliness, eudaimonic, hedonic and social well-being." Working paper. Preprint doi:10.2139/ssrn.4053961.

167 **maybe our brains:** He, Qinghua, Turel, Ofir, Brevers, Damien, and Bechara, Antoine. 2017. "Excess social media use in normal populations is associated with amygdala-striatal but not with prefrontal morphology." *Psychiatry Res. Neuroimaging* 269: 31–35.

167 **extremism and bullying:** Winter, Charlie, Neumann, Peter, Meleagrou-Hitchens, Alexander, Ranstorp, Magnus, Vidino, Lorenzo, and Fürst, Johanna. 2020. "Online extremism: Research trends in internet activism, radicalization, and counter-strategies." *Int. J. Conf. Violence* 14: 1–20.

167 **stoke real-world violence:** Siegel, Alexandra A. 2020. "Online hate speech." In *Social Media And Democracy: The State of the Field, Prospects for Reform*, edited by Nathaniel Persily and Joshua Aaron Tucker, 56–88. Cambridge University Press.

167 **that undermine democracy:** Woolley, Samuel C., and Howard, Philip N. 2018. *Computational Propaganda: Political Parties, Politicians, and Political Manipulation on Social Media*. Oxford University Press; Jamieson, Kathleen Hall. 2020. *Cyberwar: How Russian Hackers and Trolls Helped Elect a President*, rev. ed. Oxford University Press.

168 **the effects of false information:** Gonzalez, Brianna. 2023. "Neuroimaging of Political Cognition: An fMRI Study of the Encoding and Memory Retrieval of Negative Political Fake News." PhD thesis. State University of New York at Stony Brook; Moore, Adam, Hong, Sujin, and Cram, Laura. 2021. "Trust in information, political identity and the brain: An interdisciplinary fMRI study." *Philos. Trans. R. Soc. Lond. B Biol. Sci.* 376(1822): 20200140; Gordon, Andrew, Quadflieg, Susanne, Brooks, Jonathan C. W., Ecker, Ullrich K. H., and Lewandowsky, Stephan. 2019. "Keeping track of 'alternative facts': The neural correlates of processing misinformation corrections." *NeuroImage* 193: 46–56.

168 **and radicalization:** Decety, Jean, Pape, Robert, and Workman, Clifford I. 2018. "A multilevel social neuroscience perspective on radicalization and terrorism." *Soc. Neurosci.* 13(5): 511–29.

169 **for the health messages:** Jeong, Michelle, and Bae, Rosie Eungyuhl. 2018. "The effect of campaign-generated interpersonal communication on campaign-targeted health outcomes: A meta-analysis." *Health Commun.* 33(8): 988–1003.

169 **range of other messages as well:** Vosoughi, Soroush, Roy, Deb, and Aral, Sinan. 2018. "The spread of true and false news online." *Science* 359(6380): 1146–51.

169 **pitches for new shows:** Falk, Emily B., Morelli, Sylvia A., Welborn, B. Locke, Dambacher, Karl, and Lieberman, Matthew D. 2013. "Creating buzz: The neural correlates of effective message propagation." *Psychol. Sci.* 24(7): 1234–42.

173 **talk about themselves. A lot:** Naaman, Mor, Boase, Jeffrey, and Lai, Chih Hui. 2010. "Is it really about me? Message content in social awareness streams." In *Proceedings of the 2010 ACM Conference on Computer Supported Cooperative Work*, chaired by Kori Inkpen, Carl Gutwin, and John Tang.

173 **I'm not alone:** Tamir, Diana I., and Mitchell, Jason P. 2012. "Disclosing information about the self is intrinsically rewarding." *Proc. Natl. Acad. Sci. USA* 109(21): 8038–43.

173 **information about themselves with others:** Tamir and Mitchell, "Disclosing information about the self is intrinsically rewarding."

174 **disclose personal information to them:** Collins, N. L., and Miller, L. C. 1994. "Self-disclosure and liking: A meta-analytic review." *Psychol. Bull.* 116(3): 457–75.

174 **what we care about:** Berger, Jonah. 2014. "Word of mouth and interpersonal communication: A review and directions for future research." *J. Consum. Psychol.* 24(4): 586–607.

175 **share the information with others:** Cosme, Danielle, Scholz, Christin, Chan, Hang-Yee, Doré, Bruce P., Pandey, Prateekshit, et al. 2023. "Message self and social relevance increases intentions to share content: Correlational and causal evidence from six studies." *J. Exp. Psychol. Gen.* 152(1): 253–67.

175 **share the articles online:** Cosme, Danielle, Scholz, Christin, Chan, Hang-Yee, Benitez, Christian, Martin, Rebecca E., et al. 2023. "Neural and behavioral evidence that message self and social relevance motivate content sharing." Working paper. Preprint doi:10.31234/osf.io/z8946; Chan, Hang-Yee, Scholz, Christin, Cosme, Danielle, Martin, Rebecca E., Benitez, Christian, et al. 2023. "Neural signals predict information sharing across cultures." *Proc. Natl. Acad. Sci. USA* 120(44): e2313175120.

177 **Gen Z workers:** Dominauskaitė, Jurgita, Tolstych, Saulė, and Kairytė-Barkauskienė, Justė. 2024. "This is how Gen Z email sign-offs look like [*sic*], and we'll be using them." Bored Panda.

177 **smiley faces as patronizing?:** Pentelow, Orla. 2024. "You've been using the smiley face emoji all wrong." Bustle.

178 **Cyberball in our lab:** Bayer, Joseph B., Hauser, David J., Shah, Kinari M., O'Donnell, Matthew Brook, and Falk, Emily B. 2019. "Social exclusion shifts personal network scope." *Front. Psychol.* 10: 1619.

180 **people are making sharing decisions:** Baek, Elisa C., Scholz, Christin, O'Donnell, Matthew Brook, and Falk, Emily B. 2017. "The value of sharing information: A neural account of information transmission." *Psychol. Sci.* 28(7): 851–61.

180 **increase people's motivation to share:** Scholz, Christin, Baek, Elisa C., and Falk, Emily B. 2023. "Invoking self-related and social thoughts impacts online information sharing." *Soc. Cogn. Affect. Neurosci.* 18(1): nsad013.

180 **or persuade others:** Berger, "Word of mouth and interpersonal communication."

180 **social relevance's broad effects:** Cosme, Scholz, Chan, Doré, Pandey, et al., "Message self and social relevance increases intentions to share content."

180 **while their brains were scanned:** Cui, Fang, Zhong, Yijia, Feng, Chenghu, and Peng, Xiaozhe. 2022. "Anonymity in sharing morally salient news: The causal role of the temporoparietal junction." *Cereb. Cortex* 33(9): 5457–68.

181 **recommend the same thing:** Scholz, Baek, and Falk, "Invoking self-related and social thoughts impacts online information sharing"; Cascio, Christopher N., O'Donnell, Matthew Brook, Bayer, Joseph, Tinney, Francis J., Jr., and Falk, Emily B. 2015. "Neural correlates of susceptibility to group opinions in online word-of-mouth recommendations." *J. Mark. Res.* 52(4): 559–75.

182 **others do not:** Geiger, Nathaniel, and Swim, Janet K. 2016. "Climate of silence: Pluralistic ignorance as a barrier to climate change discussion." *J. Environ. Psychol.* 47: 79–90.

182 **topics like climate-related policy:** Mildenberger, Matto, and Tingley, Dustin. 2019. "Beliefs about climate beliefs: The importance of second-order opinions for climate politics." *Br. J. Polit. Sci.* 49(4): 1279–307.

182 **our motivation to do the same:** Byerly, Hilary, Balmford, Andrew, Ferraro, Paul J., Hammond Wagner, Courtney, Palchak, Elizabeth, et al. 2018. "Nudging pro-environmental behavior: Evidence and opportunities." *Front. Ecol. Environ.* 16(3): 159–68; Cialdini, Robert B., and Jacobson, Ryan P. 2021. "Influences of social norms on climate change-related behaviors." *Curr. Opin. Behav. Sci.* 42: 1–8.

182 **people's willingness to act:** Sparkman, Gregg, and Walton, Gregory M. 2017. "Dynamic norms promote sustainable behavior, even if it is counternormative." *Psychol. Sci.* 28(11): 1663–74.

185 **build even more momentum:** Fritsche, Immo, and Masson, Torsten. 2021. "Collective climate action: When do people turn into collective environmental agents?" *Curr. Opin. Psychol.* 42: 114–19.

183 **a small number of people:** Scholz, Christin, Baek, Elisa C., O'Donnell, Matthew Brook, Kim, Hyun Suk, Cappella, Joseph N., and Falk, Emily B. 2017. "A neural model of valuation and information virality." *Proc. Natl. Acad. Sci. USA* 114(11): 2881–86; Doré, Bruce P., Scholz, Christin, Baek, Elisa C., Garcia, Javier O., O'Donnell, Matthew B., et al. 2019. "Brain activity tracks population information sharing by capturing consensus judgments of value." *Cereb. Cortex* 29(7): 3102–10.

184 **the highest share counts online:** Scholz, Baek, O'Donnell, Kim, Cappella, and Falk, "A neural model of valuation and information virality."

184 **large-scale sharing than others':** Scholz, Baek, O'Donnell, Kim, Cappella, and Falk, "A neural model of valuation and information virality"; Doré, Scholz, Baek, Garcia, O'Donnell, et al., "Brain activity tracks population information sharing by capturing consensus judgments of value."

184 **less involved in that domain:** Doré, Scholz, Baek, Garcia, O'Donnell, et al., "Brain activity tracks population information sharing by capturing consensus judgments of value"; Rogers, Everett M. 1962. *Diffusion of Innovations*, 5th ed. Free Press.

186 **among thousands of people:** Chan, Scholz, Cosme, Martin, Benitez, et al., "Neural signals predict information sharing across cultures."

188 **"The nice thing":** Annenberg School for Communication. 2020. Emile: The Mission of Emile Bruneau of the Peace and Conflict Neuroscience Lab. https://www.youtube.com/watch?v=kJvfqft5v9U.

## Chapter 9: I Am the Beginning

191 **"efforts to safeguard freedom":** Ressa, Maria. 2021. "Nobel lecture." Nobel Peace Prize 2021, Oslo City Hall, Oslo, Norway.

191 **her home country, the Philippines:** Unless otherwise noted, all details about Maria's life in this chapter are taken from Ressa, Maria. 2022. *How to Stand Up to a Dictator: The Fight for Our Future*. HarperCollins.

192 **"the first page of history":** Ressa, *How to Stand Up to a Dictator*.

192 **"At the core of journalism":** Ressa, "Nobel lecture."

193 **"you are responsible":** Ressa, *How to Stand Up to a Dictator*.

194 **change how our brains work:** Kitayama, Shinobu, and Uskul, Ayse K. 2011. "Culture, mind, and the brain: Current evidence and future directions." *Annu. Rev. Psychol.* 62: 419–49.

194 **achieving larger goals together:** Kitayama and Uskul, "Culture, mind, and the brain."

194 **within the self-relevance system:** Han, Shihui, and Ma, Yina. 2014. "Cultural differences in human brain activity: A quantitative meta-analysis." *NeuroImage* 99: 293–300.

194 **how much different words described them:** Ma, Yina, Bang, Dan, Wang, Chenbo, Allen, Micah, Frith, Chris, et al. 2014. "Sociocultural patterning of neural activity during self-reflection." *Soc. Cogn. Affect. Neurosci.* 9(1): 73–80.

195 **donate money to their families differently:** Telzer, Eva H., Masten, Carrie L., Berkman, Elliot T., Lieberman, Matthew D., and Fuligni, Andrew J. 2010. "Gaining while giving: An fMRI study of the rewards of family assistance among white and Latino youth." *Soc. Neurosci.* 5(5–6): 508–18.

195 **in this case, help family:** Suárez-Orozco, Carola, and Suárez-Orozco, Marcelo M. 1995. *Transformations: Immigration, Family Life, and Achievement Motivation among Latino Adolescents.* Stanford University Press; Fuligni, Andrew J., and Pedersen, Sara. 2002. "Family obligation and the transition to young adulthood." *Dev. Psychol.* 38(5): 856–68; Hardway, Christina, and Fuligni, Andrew J. 2006. "Dimensions of family connectedness among adolescents with Mexican, Chinese, and European backgrounds." *Dev. Psychol.* 42(6): 1246–58.

196 **possible choices and priorities:** Marwick, Alice E., and Boyd, Danah. 2014. "Networked privacy: How teenagers negotiate context in social media." *New Media Soc.* 16(7): 1051–67; Goffman, Erving. 1959. *The Presentation of Self in Everyday Life.* Anchor; Butler, Judith. 1988. "Performative acts and gender constitution: An essay in phenomenology and feminist theory." *Theatre Journal* 40(4): 519–31; Hall, Stuart. 1994. "Cultural identity and diaspora." In *Colonial Discourse and Post-Colonial Theory*, edited by Patrick Williams and Laura Chrisman, 392–403. Columbia University Press.

197 **tastier than honey:** Hackel, Leor M., Coppin, Géraldine, Wohl, Michael J. A., and Van Bavel, Jay J. 2018. "From groups to grits: Social identity shapes evaluations of food pleasantness." *J. Exp. Soc. Psychol.* 74: 270–80.

197 **when people make decisions:** Tveleneva, Arina, Scholz, Christin, Yoon, Carolyn, Lieberman, Matthew D., Cooper, Nicole, et al. 2023. "The relationship between agency, communion, and neural processes associated with conforming to social influence." *Pers. Individ. Dif.* 213: 112299.

198 **cultural influence and individual identities:** Goffman, *The Presentation of Self in Everyday Life*; Butler, "Performative acts and gender constitution"; Hall, "Cultural identity and diaspora"; Uskul, Ayse K., and Oyserman, Daphna. 2010. "When message-frame fits salient cultural-frame, messages feel more persuasive." *Psychol. Health* 25(3): 321–37.

200 **engage in it themselves:** Paluck, Elizabeth Levy, Shepherd, Hana, and Aronow, Peter M. 2016. "Changing climates of conflict: A social network experiment in 56 schools." *Proc. Natl. Acad. Sci. USA* 113(3): 566–71.

201 **underlying assessment of beauty:** Amir, Ori, and Biederman, Irving. 2016. "The neural correlates of humor creativity." *Front. Hum. Neurosci.* 10: 597; Klucharev, Vasily, Hytönen, Kaisa, Rijpkema, Mark, Smidts, Ale, and Fernández, Guillén. 2009. "Reinforcement learning signal predicts social conformity." *Neuron* 61(1): 140–51; Klucharev, Vasily, Munneke, Moniek A. M., Smidts, Ale, and Fernández, Guillén. 2011. "Downregulation of the posterior medial frontal cortex prevents social conformity." *J. Neurosci.* 31(33): 11934–40; Zaki, Jamil, Schirmer, Jessica, and Mitchell, Jason P. 2011. "Social influence modulates the neural computation of value." *Psychol. Sci.* 22(7): 894–900.

201 **want to eat those foods:** Nook, Erik C., and Zaki, Jamil. 2015. "Social norms shift behavioral and neural responses to foods." *J. Cogn. Neurosci.* 27(7): 1412–26.

202 **susceptible to peer influence:** Nook and Zaki, "Social norms shift behavioral and neural responses to foods."

203 **influenced by those around us:** Yu, Hongbo, Siegel, Jenifer Z., Clithero, John A., and Crockett, Molly J. 2021. "How peer influence shapes value computation in moral decision-making." *Cognition* 211: 104641.

203 **with fully anonymous peers:** Van Hoorn, Jorien, Van Dijk, Eric, Güroğlu, Berna, and Crone, Eveline A. 2016. "Neural correlates of prosocial peer influence on public goods game donations during adolescence." *Soc. Cogn. Affect. Neurosci.* 11(6): 923–33.

204 **display it in future tweets:** Brady, William J., McLoughlin, Killian, Doan, Tuan N., and Crockett, Molly J. 2021. "How social learning amplifies moral outrage expression in online social networks." *Sci. Adv.* 7(33): abe5641.

204 **"Everything we say or do":** Ressa, *How to Stand Up to a Dictator.*

205 **watching the same speech:** Schmälzle, Ralf, Häcker, Frank E. K., Honey, Christopher J., and Hasson, Uri. 2015. "Engaged listeners: Shared neural processing of powerful political speeches." *Soc. Cogn. Affect. Neurosci.* 10(8): 1137–43; Zadbood, Asieh, Chen, Janice, Leong, Yuan Chang, Norman, Kenneth A., and Hasson, Uri. 2017. "How we transmit memories to other brains: Constructing shared neural representations via communication." *Cereb. Cortex* 27(10): 4988–5000.

205 **"Ako ang Simula":** Ressa, *How to Stand Up to a Dictator.*

205 **behave in cooperative, prosocial ways:** Peysakhovich, Alexander, and Rand, David G. 2016. "Habits of virtue: Creating norms of cooperation and defection in the laboratory." *Manage. Sci.* 62(3): 631–47.

206 **"Its systems couldn't keep up":** Ressa, *How to Stand Up to a Dictator.*

206 **how we spend our lives:** Dillard, Annie. 1989. *The Writing Life.* HarperCollins.

207 **"I didn't know":** Ressa, "Nobel lecture."

## *Epilogue*

210 **who we are:** Part of the text in this paragraph overlaps with text I wrote that appears in Ben Shattuck's *Six Walks.*

# INDEX

Page numbers in *italics* refer to illustrations. Page numbers after 219 refer to notes.